AF873472

The Human Factor in Project Management

Best Practices and Advances in Program Management Series

Series Editor
Ginger Levin

RECENTLY PUBLISHED TITLES

PgMP® Exam Test Preparation: Test Questions, Practice Tests, and Simulated Exams
Ginger Levin

Managing Complex Construction Projects: A Systems Approach
John K. Briesemeister

Managing Project Competence: The Lemon and the Loop
Rolf Medina

The Human Change Management Body of Knowledge (HCMBOK®), Third Edition
Vicente Goncalves and Carla Campos

Creating a Greater Whole: A Project Manager's Guide to Becoming a Leader
Susan G. Schwartz

Project Management beyond Waterfall and Agile
Mounir Ajam

Realizing Strategy through Projects: The Executive's Guide
Carl Marnewick

PMI-PBA® Exam Practice Test and Study Guide
Brian Williamson

Earned Benefit Program Management: Aligning, Realizing, and Sustaining Strategy
Crispin Piney

The Entrepreneurial Project Manager
Chris Cook

Leading and Motivating Global Teams: Integrating Offshore Centers and the Head Office
Vimal Kumar Khanna

Project and Program Turnaround
Thomas Pavelko

The Human Factor in Project Management

Denise Thompson, MBA, PMP, CSM

CRC Press
Taylor & Francis Group
Boca Raton London New York

CRC Press is an imprint of the
Taylor & Francis Group, an **informa** business
AN AUERBACH BOOK

Cover illustration and drawings by Jim Kangas. Reprinted with permission.

"DiSC®" is a registered trademark of John Wiley & Sons, Inc.

"Excel®," "Microsoft PowerPoint®," and "Microsoft® Project Professional" are registered trademarks of the Microsoft Corporation.

"MBTI®" is a registered trademark of the The Myers & Briggs Foundation in the United States and other countries.

"PMI® Global Congress," "*PMBOK® Guide*," and "*Pulse of the Profession®*" are registered marks of the Project Management Institute, Inc., which is registered in the United States and other nations.

"Post-it®" is a registered trademark of the 3M Company.

CRC Press
Taylor & Francis Group
6000 Broken Sound Parkway NW, Suite 300
Boca Raton, FL 33487-2742

© 2019 by Taylor & Francis Group, LLC
CRC Press is an imprint of Taylor & Francis Group, an Informa business

No claim to original U.S. Government works

International Standard Book Number-13: 978-1-138-06419-5 (Hardback)

This book contains information obtained from authentic and highly regarded sources. Reasonable efforts have been made to publish reliable data and information, but the author and publisher cannot assume responsibility for the validity of all materials or the consequences of their use. The authors and publishers have attempted to trace the copyright holders of all material reproduced in this publication and apologize to copyright holders if permission to publish in this form has not been obtained. If any copyright material has not been acknowledged please write and let us know so we may rectify in any future reprint.

Except as permitted under U.S. Copyright Law, no part of this book may be reprinted, reproduced, transmitted, or utilized in any form by any electronic, mechanical, or other means, now known or hereafter invented, including photocopying, microfilming, and recording, or in any information storage or retrieval system, without written permission from the publishers.

For permission to photocopy or use material electronically from this work, please access www.copyright.com (http://www.copyright.com/) or contact the Copyright Clearance Center, Inc. (CCC), 222 Rosewood Drive, Danvers, MA 01923, 978-750-8400. CCC is a not-for-profit organization that provides licenses and registration for a variety of users. For organizations that have been granted a photocopy license by the CCC, a separate system of payment has been arranged.

Trademark Notice: Product or corporate names may be trademarks or registered trademarks, and are used only for identification and explanation without intent to infringe.

Visit the Taylor & Francis Web site at
http://www.taylorandfrancis.com

and the CRC Press Web site at
http://www.crcpress.com

Dedication

To the amazing project teams I have had the privilege to serve over the last two decades. Through my service to you, I have learned volumes about success, failure, leadership, and character. Thank you for your dedication and commitment to making a difference in this world.

Contents

Foreword

This book will touch your life, transform your approach to project management, and revolutionize your relationships with your team and customers. In this masterpiece, Denise Thompson brilliantly reveals the success key that opens the door to helping you provide more value to people's lives and build your legacy that lives long after your project ends. Follow Denise Thompson's teachings—not only will you lead more successfully—you'll LOVE what you're doing even more! If you're a project manager, this book is one of those you'll not only want everyone on your project team to read, but you'll most likely want to make it a study. It's that good!

— Jennifer Bridges
CEO, PDUs2Go.com
Author, *Optimize Your Thinking: How to Unlock Your Performance Potential*

Preface

One day I woke up to the shock that a wave of change has hit the profession of project management and we are not ready. In fact, most of us don't see the storm coming and we don't have a life vest to survive in the rough seas ahead.

Over the past five years, I have heard countless stories of the tumultuous and painful end of a Project Management Office (PMO). While this is our current reality, I continue to find myself at local and national Project Management Institute (PMI®) meetings that focus on endless debates centered around the methodology, framework, or tool that might best serve our projects. As I listen to the debates, my anxiety and frustration builds as I wonder why are we not focusing all our efforts on redefining our value in a way that does not rely on factors we cannot possibly control, or sometimes predict. While the debates continue among us, precious opportunities to work together to redefine the value of project management and assure a sustainable future for the industry are lost.

As time marches on, consultants and innovation groups are filling the gaps between the project value that is promised and what is actually delivered. Today, you can find any variation of new tools and methods that claim to increase project, program, and portfolio success by simply checking out YouTube or Google®. Still, you won't find a magic formula there that provides consistent project success. So where is this elusive magic formula for project success? And without this magic, how do we define our value?

For over half a century now, the value of having a project manager has been proposed as one that delivers project success through control of the Iron Triangle. The Iron Triangle, also known as the Triple Constraint, is represented as the forces of time, scope (the work included in a project), and cost (including human resources and funds). Consistent average project failure rates that exceed 50 percent have taught everyone affected by the project (defined as stakeholders) that project management cannot deliver as promised. In response, more and more organizations are clamoring to find new tools and techniques that will manage the tumultuous forces of the Iron Triangle. More organizations are turning away from formal project management and using a more ad-hoc approach by engaging general staff with little to no project expertise to manage projects. At the same time, organizations continue to pour billions of dollars into projects worldwide every year. Even with so much at stake, the answer to what creates consistent project success eludes us. Why? Maybe it is because we are looking for the answer in the wrong places. Where is the right place? A good look in the mirror will start us on the right path to find the answers.

The harsh reality is that the organizations we work for do not really care how we manage project scope, people (human resource management), project funds, and schedules, as long as the project manager and the project team meet the needs and expectations of the organization and project stakeholders. In fact, I find that executives often require only from project management what our own profession has taught them to expect. The trouble is that expectations of controlling the Iron Triangle are not realistic in a fast-changing environment, because the sides of the Triangle are anything but Iron. Today, the Iron Triangle is more like a wet noodle that proves impossible to fix in rapidly changing environments.

Rapid changes demand a fully integrated leadership approach in project, program, and portfolio management, yet staunch resistance to change still exists within the profession of project management. This resistance to change is perplexing given the overwhelming evidence all around us that change and acceptance of change are inevitable and required for survival. Perhaps this resistance explains why the average life span of a project manager and a project management office is less than 10 years.

Whether you've been involved in projects for two months or 20 years, you have probably noticed the same themes reoccurring. Sponsors, executives, and project managers tend to focus on tools to help them manage the tough issues that threaten project success. This is a curious approach, because the real problems that threaten project success are rooted in the human factors that are ever-present in our work and personal lives. These factors cannot be managed through a spreadsheet, a team-building exercise, or a requirement. Human factors are influenced and managed through human behaviors.

The Human Factor in Project Management challenges you to go on a journey of self-discovery through a mélange of personal and professional experiences. In each story, you are asked to reflect on your own feelings, how they drive your actions and behaviors, and the results those behaviors are likely to drive in others. Whether you serve as a portfolio manager, program manager, sponsor, or stakeholder, you will be challenged to discover your own barriers to change and achieving a greater level of understanding.

Along the way, you will take a fresh look at project and professional failures in history, learn the value of perseverance, redefine success, and see the importance of leadership. Together, we will look at the value of waterfall and agile methods and examine the debate about which method is best. Our journey will explore the differences between waterfall schedules and Kanban boards and learn the impacts of both. At every turn, you will be challenged to dig in and find your courage to lead.

The obvious problem in our industry is the absence of a relevant value proposition. But if our value cannot be based on the forces of the Iron Triangle, what is our real value?

The Human Factor in Project Management challenges every one of us involved in projects to redefine the value we bring to our organizations through our leadership. Throughout this book, leadership is seen as synonymous with project management. This type of leadership is a unique mix of coaching, mentoring, teaching, and managing that is required to navigate the sea of human project variables. To become this type of leader requires hard work, self-discovery, and change. Will you choose to embark on the journey?

The reward for taking the journey is the promise that you will create the opportunity to secure your own relevance, value, and future in project management. On the other hand, if you reject change and choose the status quo, you will guarantee your inevitable removal from the future landscape of project management.

Here is my challenge to you: Embark on the journey. Learn to become a true leader and driver of change for your team, your organization, and the profession of project management to redesign the value proposition for project management. Design a new value proposition as a leader who creates project success through courageous, genuine leadership. This is important because what will determine our future as project managers and the relevance of our industry cannot be found in a spreadsheet or a methodology. The answer is right there in the mirror. Do you have the courage to take a hard look? Do you have the courage to lead the human factor in project management?

Acknowledgments

It's is with a debt of gratitude that I thank Dr. Ginger Levin for giving me the opportunity to share my passion for project leadership with the world. Without Ginger's support and leadership, the lessons I've learned about leadership would remain as faint messages that exist only in my mind.

Ginger is a tireless warrior in the world of project leadership. Through her example, we all can learn to coach, mentor, and lead those who are dedicated to creating a better tomorrow.

About the Author

What makes this author and this book different is found in the relevant experience Denise Thompson brings to the page as a working project management professional (PMP) and certified Scrum master (CSM) and MBA. Fluent in the various practices of agile, waterfall, and the countless variations that fall in between the better-known methodologies, Denise has developed project leadership through a myriad of personal and professional experiences.

Throughout a wide array of diverse professional and personal experiences around the world, which have taken Denise into the Arctic, the desert, and urban and rural jungles in corporate, cooperative, and nonprofit environments, Denise has been continually challenged to define and redefine herself and her approach in order to adapt to new environments, requirements, and stakeholders. Her passion for project management is matched only by her passion for relevant leadership. As a servant to teams and organizations, Denise is an agent for change and an advocate for the development of a sustainable future for the profession of project management. There is nothing to sell here except the revolutionary idea that every project manager has the ability to earn project success and design the future survival of project management.

Chapter 1

Are You a Stick in the Mud, a Cog in the Wheel, or Skids on Rails?

A journey of a thousand miles begins with a single step.

— Chinese proverb

© Jim Kangas

There are many factors to consider in the world of project management. One of the most important is the one that is often most overlooked. Why? Because to examine this factor and create a formula for your success as a leader and a project manager requires the hard work of embarking on a learning journey that demands self-reflection, specific leadership behaviors, and skills in the ability to embrace change and adapt quickly. Like anything in life, if you want to earn genuine success and the respect of those around you, you are going to have to work hard. This book is not the answer to your success. It is simply the key to learning how you can earn it. As you read, you will note that *project manager* and *leader* are treated as synonymous terms. It is as if a project manager is a specific type of leader, like chocolate is a specific type of ice cream.

As we embark together on this journey, think of yourself as a leader of people and not someone whose success is predicated on a pretty Microsoft PowerPoint® or impressive Gantt chart. Those are the tools of the trade. Tools, no matter how effective, eventually change, become obsolete, or useless. It is the human factor in project management that drives success and earns results, not the tools. It is, in fact, *you!* Considering that companies routinely dedicate 40 percent or more of resources to project and information technology (IT)–based initiatives and are wasting an average of $97 million USD for every $1 billion USD invested in projects (Project Management Institute [PMI], 2017b), you may be the most powerful asset within the organization to produce three key deliverables: earning value, building teams, and project performance. Do you think you can accomplish all three of these deliverables? Let us find out.

1.1 What Really Controls a Project

Since the time of the pyramids, project management has based its value on the ability to complete project deliverables on time, on budget, and within scope. The relationship among these three factors is called the Triple Constraint or the Iron Triangle. The skill of the project manager in wrangling the infamous Iron Triangle, which represents project resources, schedule, and scope, has focused primarily on the prescribed use of an array of tools, processes, and procedures. In the past, the concept was the belief that a seasoned project management professional had the mastery of tools which inevitability control unyielding schedules, manage escalating budgets and uncertain resources, while holding a firm grip on project scope. In fact, as new project managers become seasoned, they learn that the tools and processes which provide aid to understand, mitigate, and reduce risk to the factors representing the Iron Triangle do not control anything. Why? People and behaviors control events and outcomes, not tools and process. Let's look at an example.

Think of people in your life who are notorious for taking their time to get somewhere. They are not terribly concerned about arriving at any event on time or even early. You, on the other hand, are hell-bent on always arriving early. Imagine you and a friend are driving together to an event today. You arrive early to pick up your friend Sue. Sue is well known for being late. You wait in the car, texting Sue of your arrival. You are anxious, trying to figure out a way to prod Sue to hurry. After five minutes, you become increasingly concerned that you both will arrive late to the event. You decide to walk up to the door to urge Sue out of the house and hurry to the car so you both can leave for the event on time. Instead, Sue invites you into her house. After another five minutes, you state your concerns and suggest you may want to drive separately. Sue responds, "I'll only be a couple of minutes, hang on." In the end, you both arrive together late to the event. You are frustrated and Sue senses this, which sets a negative tone between you for the rest of the evening. Reflect for a moment on how successful you were in preventing the very outcome that you did not want. How did you change Sue's behavior? What can you do to improve the outcome next time, so you both have a better experience and arrive on time?

The bottom line is that you can do nothing on your own to improve or change a future outcome that is reliant on the behavior and actions of other people. It is your ability to model the way and inspire a change in the other party's behavior that will inspire change in the future. It is Sue's willingness to change and be receptive to adopt new behaviors that will allow you both to arrive on time in the future. By now you are probably asking, "What does this have to do with project management?" Let me tell you. Everything.

Today, project managers and stakeholders alike are wising up to the fact that the Iron Triangle is an ocean of ever-changing variables that yield very few predicable outcomes. In business environments, which endure buyouts, takeovers, mergers, acquisitions, and more change than we can track, all occurring at the speed of light, the adage that we can only control our own actions and reactions has new meaning and importance.

1.2 What Affects the Outcomes

It is more important than ever to understand that the success of project managers depends on their ability to wield influence and guide outcomes toward a result that stakeholders value. What does that really mean? It means that when you get down to how stakeholders define your success as a project manager, your behavior, the perception of that behavior, and your intent is how you are measured as a professional.

1.3 It's About Relationships

To understand more about how the ability to exhibit successfully the behaviors that develop solid trust and relationships with stakeholders drives positive outcomes, think about Sam. Sam is a project manager who is responsible for a project that is now experiencing a slipped schedule, a blown budget, a dysfunctional team, and deteriorating confidence. Through it all, Sam remains in good standing with the project sponsor and steering committee. In this turbulent environment Sam is known as the calm through the storm by everyone around him. He refrains from placing blame, he is quick to own his mistakes, he openly seeks input, and he respects his team and colleagues. Sam is eager to support the team and is proactive in removing barriers to make the team's work efficient. He is reluctant to take center stage or give credit to himself. For Sam, it is all about the team. Sam is careful to exhibit his character through consistent, reliable actions and is known for keeping his word. Many might define Sam's behavior as playing politics or applying political savvy. Yet, if you ask people who work with Sam, they simply say he can be trusted to get the job done in a way that promotes solid relationships that remain long after the project is closed. Sam has earned his good standing by working to create meaningful, genuine relationships built on trust. Because of these solid relationships, the trust between Sam, the sponsor, the team, and the steering committee remains stable through turbulent times. The foundation of trust allows the group to be supportive and provide what Sam and the team needs to turn things around and get the project back on track to deliver measurable value. In the end, Sam and the team remain cohesive, and they continue to be supported by the steering committee and sponsor even though they deliver two years late and are $1 million USD over budget. Despite missing the original budget and schedule targets, the project is defined as a tremendous success by the organization. How could this be? Certainly, the delay and increase in budget reduced whatever return on investment (ROI) was estimated for the project. There may also be opportunity costs, realized risk, issues, and other consequences that impact the organization financially because of the project budget and schedule overruns. To make the scenario even more confusing, the outcome is not always this pleasant for project managers who find themselves struggling with a dysfunctional team, budget overages, and slipped schedules. So why, given the same project failure, would the outcome be different for one project manager over another?

1.4 Human Behavior—The Real Driver of Projects

To answer this question, let us examine the definition of a project as defined by the Project Management Institute (PMI) in *A Guide to the Project Management*

Body of Knowledge (*PMBOK® Guide*), Sixth Edition, as "a temporary endeavor to create a unique product, service, or result" (PMI, 2017a). By that very definition, project management is a journey into the unknown. As project management professionals, we are leading the organization to complete an effort that has not been achieved before exactly as the project scope outlines it for that particular effort. Just like walking alone down a dark alley in an unfamiliar city, going on a project journey can ignite in us some of the same primal fears that the cave men experienced when exploring a dark unfamiliar cave. Science does not have a clear definition of what spawns these fears, because what triggers the sensation of fear is different for everyone based on the individual's past experiences and perceptions of the past, present, and future. To understand this fact is important, because fear is the undercurrent that drives your choices and behaviors as well as the choices and behaviors of everyone around you.

Maslow's hierarchy of needs, developed in 1943, established five psychological needs that spark a series of reactions in humans when they are threatened. The base of these needs, described as a pyramid, is categorized as our basic needs. These are the need to feed, clothe, and shelter ourselves and our families, and the need to feel safe and secure (McLeod, 2017). Described in this way, for many of us, these basic needs can quite literally be connected to the need to keep our job. Reactions to these basic needs are described by Maslow as some of the strongest in the human species, born from our primal instinct to survive. Does thinking about this put a new light on boardroom antics for you? If it doesn't, maybe it should.

What is really playing out in the boardroom when things get tough? Chances are, we are witnessing a reaction to fear. We might be reacting to the fear that one of our needs will not be met and the fear that our self-preservation in the form of our job is in jeopardy. When you reflect on behavior in the boardroom, or in any meeting at work where the stakes are high, how often do you see what could be categorized as a fight-or-flight behavioral response by those involved in the meeting? Establishing blame, preparing political defenses, framing messages and facts to create a desired perception of the information, or shutting down, denial, and the elusive "I do not recall," "I don't understand," or "Let me get back to you on that" response can signal a flight from a tough situation at hand. Behind the scenes, establishing an offense, gathering allies, making quid-pro-quo deals, are all actions we might take to prepare for the fight to come and assure our self-preservation, self-actualization, and self-promotion. These are strategic moves, even politically savvy moves, that we carefully orchestrate to increase the opportunity to build our success and that of our organization and to reduce the risk of failure as we perceive it in the situation. Also, our own perception is not the only factor at play. The perceptions, actions, and reactions of others can have an effect on the outcome of any situation. If you ignore this reality, you are ignoring the real story that is unfolding. This is the story that will be judged. This is the story that will define how you are perceived in your organization.

1.5 The Power of Group Dynamics

Group dynamics is defined as several individuals who come together to accomplish a task or goal and refers to the attitudinal and behavioral characteristics of a group. Group dynamics focuses on how groups form, their structure and process, and how they function. In addition, group dynamics can increase stress and anxiety in individuals, spreading negative emotions throughout the group, whose members often translate these negative emotions into negative and harmful behavior. This bad behavior essentially amplifies the effect of actions and reactions in stressful situations, creating a concoction of human emotions and reactions that can be volatile and impossible to navigate and predict. To understand better the impact of group dynamics when people are affected by negative stress, anger, and fear, let's examine a few events in recent history. These events left indelible marks on our entire nation in the 1990s.

On April 29, 1992, a court in California handed down a "not guilty" verdict in the case of four white police officers who were accused of severely beating an African American man named Rodney King in Los Angeles the previous year (History.com Staff, 2009). Almost immediately, Los Angeles fell under siege from its own citizens. What started as a small protest soon became an angry mob. Five days later, property damages inflicted by the mob were estimated at $1 billion USD. Fifty people had been killed, more than 2000 injured, and 12,000 people had been arrested, resulting in what some say was the worst incident of mob violence in U.S. history (CNN Library, 2013). Afterwards, some of those involved in the violence stated they felt they were victims of "mob mentality." Mob mentality is described by Flow Psychology as a state when human actions and reactions are influenced by the peers around them. It leads to others adopting behaviors from the group that they might not normally exhibit. The influence can be profound (Flow Psychology, 2014).

Throughout history, there have been many examples of human behavior under stress. The result is not always negative. Think of the attacks on U.S. soil on September 11, 2001. Americans are very familiar with the outpouring of accounts of people who reacted with courage and valor to the events that occurred on 9/11 to defend and protect the lives of others. The account of Flight 93 is one example.

On the morning of September 11, 2001, United Airlines Flight 93 departed from Newark, New Jersey, heading to San Diego, California. The plane carried 7 crew members, 33 passengers, and 4 hijackers. Forty minutes into the flight, the hijackers made their move to take control of the plane. The passengers gathered in the back of the plane, where they voted to plan and execute a counterattack against the terrorists. At approximately 10:03 a.m., the plane was crashed by the terrorists just east of Pittsburgh, Pennsylvania, in an effort, it is believed, to prevent the passengers from regaining control of the plane. The courageous actions of the passengers foiled the terrorists' plan to crash the plane into the White House or the Capitol Building in Washington, DC, which was believed to be the ultimate target. If the plan had succeeded, countless more lives would have been lost, significantly increasing the devastation of the 9/11 attacks (National Park Service, 2017).

The historical account of Flight 93 demonstrates how a group of strangers experiencing a direct threat to their lives and the safety of others joined together to overcome their own irrational, emotionally reactive responses to act cohesively, collaborating to overthrow the hijackers, at the ultimate cost of their own lives. The story of Flight 93 is an account of an act of heroism that literally saved hundreds of lives.

1.6 Fear Is the Greatest Cost of Project Failure

Are you wondering how these examples relate to human behavior on projects or in the boardroom? Maybe you are thinking these historical lessons are too extreme to apply to project management? In this chapter, I will present some historical facts that might just change your mind. The point we need to remember is that group dynamics can negatively affect outcomes whether we are on a plane that has been hijacked or we are dealing with a million-dollar project budget that is in jeopardy of serious cost overruns. To demonstrate these dynamics on projects, let us look at a few more examples from history.

The Panama Canal, known from its inception as an engineering feat of the time, started its project journey with an initial investment of $120 million USD. In 1880, this was a substantial sum of money. To compound investor fears, the building of the canal was expected to be one of the most risky, expensive, and high-tech projects of its day (Global Security, 2011). From the project kickoff, anxiety over costs, labor unreliability, injury and death, difficult terrain, and other known and unknown risks was exacerbated by the behavior of the humans involved. At first glance at this story in history, the causes of project failure seem clear and out of the project manager's control, so maybe we should look a little closer.

Ferdinand de Lesseps was still reaping the benefits he had earned from his success as developer of the Suez Canal, and glowed with a confidence that some would say later was arrogance. In 1879, Ferdinand proposed building a canal across the Isthmus of Panama. Based on the financial success of the Suez Canal in Egypt just 10 years earlier, investors were eager to support the project. The Compagnie Universelle du Canal Interoceanique was incorporated under French law on March 3, 1881, and de Lesseps hired 500 young engineers for the project; he predicted that construction would be complete in three years. At the time, no one could have imagined that none of the 500 engineers would return alive from Panama. Twelve years after the start of the project, costs had doubled above original projections, and although survey work had been done, the canal was yet to be built. Although initially the press focused on the colossal construction failures and worker deaths caused by illness and injuries as the reason for project failures, behind the scenes, corruption, misinformation, misappropriation of funds, and fraud soon came under the scrutiny of the law. Eventually, the corruption on the project surfaced and became known as the primary cause of failure of the biggest project since the Suez Canal. These failures, once made public, almost overnight destroyed the career of the man who just years before had been known as one of the best project managers of his time (Global Security, 2011).

The behavior that contributed to this sad story included de Lesseps' failure to report on the actual project cost and schedule overruns, his failure to initiate and implement changes, and his failure to report accurately these changes, costs, and the impacts to investors. When de Lesseps' practices were reviewed, it was found there were essentially very few records of issue management or change management occurring on the project. Let us assume that de Lesseps himself was not aiding in the fraud and bribes of the press and government officials that effectively re-routed funds intended for construction. The amount of funds pocketed by the officials of the Compagnie Universelle du Canal Interoceanique is not known, but it is clear that de Lesseps, as the project manager and president of the company, took extreme actions to cover up bad news. Fear, maybe arrogance, and perhaps greed were drivers of de Lesseps' behavior. He did not choose to reconsider his plan for the project to allow for the construction of smaller canals, which might have helped move things forward. He did not manage investor fear or quickly address human resource losses on the project, which were attributed to the harsh environmental conditions. It is known that de Lesseps did not trust the engineer experts he so eagerly enlisted at the start of the project, and he thought he could correct project failures and ultimately create success for the endeavor before the stakeholders and the press were the wiser. In the end, de Lesseps' Panama Canal was never finished and only one-third of the funds for the project were used for the actual construction, leading to the conviction of de Lesseps and his company's officials for fraud and embezzlement. Ferdinand de Lesseps ultimately died disgraced and impoverished. In 1904, the United States bought the rights to the Panama Canal project. By its completion in 1914, the project was the most expensive in U.S. history, costing over $375 million USD (Federer, 2016).

Is the story of de Lesseps' canal so different from project management today? Be honest: Has the thought ever crossed your mind not to share bad news with stakeholders? In my 20-plus years of experience, the strategy to initially withhold news that might spark negative reactions and even ignite fears from stakeholders is almost always at least a consideration on a project, whether it comes from stakeholders, sponsors, project managers, or the C-suite. Is this ethical behavior? No. Is this reality? Yes. So, what do you do when you are facing the choice to share bad news or wait to see if the bad news blows over?

What defines you as a project manager and ultimately defines your career is your ability to embrace new and different behavior and adapt to become who you want to be in the toughest moments. I think you know what moments we are talking about. It is those moments when you can smell the fear in the room—those moments when the greatest threat to you and the project is a negative reaction that can catch like wildfire in a group of stakeholders (group behavior) and manifest itself in any number of forms of negative group-think, or even mob mentality. Your ability to look beyond the moment and into the future to drive positive outcomes will give you the ability, not to control, but to navigate the negative behavior and fear that poses the greatest threat to any project. Remember, the one thing you truly control is also your greatest asset to success as a project manager and leader. That one thing is your own behavior. It is undoubtedly your greatest asset and your most dangerous saboteur.

1.7 A Look into the Mirror

The first step to controlling your behavior and adapting to the situation is to understand yourself. This means taking time to reflect. Does this sound easy to you? How much time do you spend digging into what triggers your fears, your anger, and the experiences that form them? Most of us might try to avoid this exercise. Why? Speaking from my own experience, self-reflection can be difficult. We would probably agree that it is certainly more challenging to look deep inside yourself than to spend time blaming external factors for your actions and reactions. When you are asked why the project is in poor health, how many times have you used the excuse, "The team is not performing," or "The sponsor keeps changing the scope"? Next time you are tempted to use this as your scapegoat for mediocre performance, go to the mirror, look at your reflection, and examine your own behavior and the outcomes you have been able to achieve. If your behavior is inspiring the team to perform, the conversation you have with your reflection will be short and sweet. If the team is not performing, your behavior might be the culprit. Do you have the guts to take a hard look at your own reflection?

Self-reflection is difficult because it's scary. After all, digging too deep might uncover things about yourself that you would rather not face. Maybe that is why it is much more palatable to dive into a budget or a Gantt chart to find the answers that will help you guide and inspire stakeholder behavior that will support project success. The problem with this approach is that you will not find the answers you need in those tools. If we can find the courage to focus on a journey of self-discovery, we can give ourselves the opportunity to unleash our true ability to influence the behavior of others and adopt behaviors that are the foundation of project performance and, ultimately, success. We are giving ourselves the opportunity to truly become leaders. Ready? Let's start digging!

Remember the definition of project management as an endeavor to create something new, something unique. The new and unique piece of that definition is a crucial clue to an important skill set that project managers need to develop so they can lead the team through a project journey. Going somewhere you have never been before requires navigating through the unknown, flexing to changes and adapting. Is this a skill that is part of your tool box?

1.8 The Story Reveals Your Character

Imagine you are going on a train trip through France for the first time. You have read travel magazines and did as much research as possible to prepare yourself for the journey. You are responsible for bringing five people along on this trip. The goal is to get from Nice to Lille, France. You know it is about 720 miles or 10 hours' drive on the Autoroute. Your mission is to get there with your group in 9 hours or less. After a bit of research, you discover the solution to fast-track the journey. It is easy, you are thinking, just take the fast train, the TGV. The route is considerably faster and should get you to your destination in 8 hours. So off you go with your group map in hand and Google at the ready on your smart phone. It is early morning on a beautiful fall day. There is a cool breeze and the sun is shining. You are standing at the ticket counter now, preparing to purchase the tickets for the group and embark on your journey. How do you feel? Anxious, excited, scared, all the above? Are "what if" scenarios running through your mind? Do you have a backpack full of emergency gear, or do you simply carry your credit card and your jacket? What is the weather like? Do you notice? Have you researched the weather, or learned the language to prepare for this trip, or are you content to use nonverbal cues and prefer to bet that you will probably find someone who speaks English if needed? Maybe you are wishing you were back at home, back in a place that is familiar and you were free to leave this place that is so unfamiliar, different, and, well, just different.

You have arranged for the group to sit together to reduce the risk that anyone would stray away and lose their way. As you settle in your seat with the group at the front of the train car, another passenger walks up with her ticket, stops in the aisle by your seat, and signals to the conductor. Although you cannot quite understand what they are saying in French, you have a hunch that there must be some confusion about the seat assignments. After a couple of minutes of discussion, the conductor walks over to you and signals for you to get up and move to an empty seat at the back of the car. How do you feel? What do you do now?

What seems like minutes later, you are awakened by the sudden stop of the train. When you open your eyes, people are chatting, the prompter screens are black, and the conductor is ushering people out of their seats and into the aisle to the brisk air outside. Once off the train, you can see passengers getting off all the cars on the train, stepping outside into the cold. You hurry to catch up to your group, huddled together alongside the tracks. You view the train and the vast open, desert-like landscape. No road or station is in sight. What is going on, you ask the others? "I'm not sure," someone in the group says, "but I think I heard the conductor say 'la gréve'—that's the word for strike. If the train's employees are on strike, this could mean the end of the ride for us." A few minutes later the train pulls away. Some of the passengers yell for the train to stop, to no avail. As the train rolls away from the passengers, you are able to view the other side of the tracks, where luggage is piled in a heap on the ground. Passengers hurry to the heap, only to struggle to find their bags. With bags in hand, most of the passengers head toward the west. Your group discusses this and decides to follow the other passengers, even though you have no idea what lies ahead. Cell phone coverage and Google are unavailable. After 20 minutes and no hint of civilization, two members of the group decide the best strategy is to stay put with their suitcases and wait for help. The air is cooler now, and the sun is at 4 o'clock. You know as the sun sets it will get cold, and with no road nearby it is unlikely that help will arrive. If in fact the railway workers are on strike, no other trains may come by for days. As the group leader, you feel responsible and do not want to leave anyone behind, so what do you do now? The group is depending on you, yet they are reluctant to venture on a direction with no clue of how long or how far they must travel to get back to civilization.

What happens next will reveal your true character and capability as a leader, demonstrating whether you are *change agent,* a leader who has the ability to adapt to and embrace new and unforeseen events. A change agent is also someone who can articulate a clear vision and direction that helps others align to that vision. A change agent engages in active, empathetic listening to inspire others to trust, all while creating a sense of urgency for change, empowering others to act, leading the way with confidence, yet remaining humble while encouraging people to work together. Does this sound like you?

1.8.1 Stick in the Mud

Maybe our story will reveal your character as a "stick in the mud," resisting change, unwilling to go into a new situation or risk the unknown, standing firm in the status quo and your own familiar place. After all, you have always been there, it is comfy, so why not?

1.8.2 Cog in the Wheel

If those descriptions do not fit your style, you might be more like a "cog in the wheel." This means you will do something if everyone else is already in line. You will follow the flow and direction of the group, not willing to step out of bounds or lead the way. After all, it is much easier to follow the flock than to forge the trail for others to follow.

1.8.3 Skids on Rails

Finally, your actions in our story might show that you resemble more of a "skids on rails" personality type. There are no brakes or direction for you; you will go with whatever direction seems right at the time, without fact or data to draw from, rapidly embracing anything new, rolling over anything in your way in the process.

1.8.4 Why Are These Characterizations Important?

These characterizations and their outcomes are not a judgment of character; rather, they provide information to help you evaluate who you really are and how you can adopt the characteristics you need to succeed.

Ancient Chinese wisdom states succinctly what we are talking about here: "Be careful of your thoughts, for your thoughts become your words. Be careful of your words, for your words become your actions. Be careful of your actions, for your actions become your habits. Be careful of your habits, for your habits become your character. Be careful of your character, for your character becomes your destiny."

Project managers, in the role of leading others, share their destiny with the team during the life of the project, and often well into the future. Considering this, your choices can have a profound impact on your career, the career of others, and your organization.

Choices we make in either our professional or personal life are truly a reflection of our character and our capability to adapt and embrace change.

Change is a constant in today's business environment. Project managers have the responsibility of leading and embracing change. It is intrinsic in what we do as we lead subject-matter experts to develop unique products, deliverables, technology, architecture, services, and tools.

1.9 It's All About Choices. What Will Yours Be?

Let's go back to the moment of choice on that ill-fated train trip in France. Take a moment to look at the following options and play them out in your mind. What will you choose?

A. Agree with two of the team members to encourage the others to stay put.
B. Threaten the two members who are lagging behind and push the group to follow you to the east.
C. Go west with the rest of the passengers, encouraging the two teammates lagging behind to join you.

A new endeavor is a journey that stirs up emotions regardless of whether you are in the midst of a business endeavor or a personal adventure. How many times have you seen tempers flare in a tough business environment? We are human, and the drive that empowers us is emotion. Yet too often we ignore human behavior as a critical component of project management.

The ability to recognize what you are doing that is not working and adapt to embrace change is crucial to project success and your ability to create a positive experience for stakeholders, remembering that the success of the project is often tied to the human experience rather than how well we manage the ever-evolving sides of the Iron Triangle.

Referring to that train trip in France, let's examine the choices a bit deeper and consider what they might reveal about your character.

Option A. Agree with two of the team members to encourage the others to stay put.

If this option suits you, you might be thinking this seems like the safest option available at the moment. Certainly, waiting out the night in the same spot is better than going into the unknown? Besides, just because the other passengers head to the west does not mean that this is the one direction that will lead to a road or a rescue. Who knows what the right move is, anyway? None of you have been here before. You are in a strange country, with no cell service and no maps. The fact is, none of you have experienced this before, and you don't even speak the language. It's a downright scary situation. Anything could happen. Isn't it better to sit tight than to risk getting lost moving forward or potentially traveling away from help rather than toward help? You know that, without more information, either moving forward or staying put is a calculated risk. The larger group is fearful of staying put for the night. Some in the group don't have warm enough clothes for the cold that is coming all too soon. What about food and water? Some people in the group have a few provisions. You never imagined this would happen, so you find yourself unprepared, with only a jacket, a bottle of water, and an energy bar. You are thinking this is the worst experience you have been through and it's just beginning. The fear and frustration in the group is growing, and assumptions on outcomes that might result from choosing to stay put or move forward are flying everywhere. The group is divided. You do your best to convince your team to stay calm and make the best of a bad situation. Ultimately, the loudest voices in the group prevail, but they too are divided. You are unsuccessful, and although you try your best, the team continues to move forward while you remain with the two who chose to stay behind. Ultimately, you feel this is a safer bet even though the group is now split and the three of you will face the night alone in the cold, unfamiliar countryside.

Does this sound like a choice you would make? If so, you might be a "stick in the mud," resisting anything new, anything uncertain, or anything that requires change. When the environment and the future is uncertain, you prefer to stay put.

Option B. Threaten the two members who are lagging behind and push the group to follow you to the east.

You figure anybody's guess is a good guess at this point. No one has information on which to make a decision anyway. It is a 50/50 chance either way by your calculation. Right now, you feel that any movement is the right direction, just as long as you are moving forward. If the group wants to follow along, that's their choice. You really are not interested in trying to convince anyone of anything. Frankly, you feel it's not worth the energy. So, you have decided to go east without data to support the choice—after all, it makes just as much sense as going in any other direction. Besides, with the others going west and your group going east, maybe this way someone in one of the two groups will find a road and get help for the rest of the passengers. After other members in the group do their best to convince you to stay with the team, you move forward to the east, leaving the group behind to do what they will. What else can you do; you do not control their behavior anyway, right? If this scenario sounds like a choice you would make, you might be a "skids on rails," moving forward without information because moving forward feels like progress, regardless of the cost.

Option C. Go west with the rest of the group, encouraging the two lagging behind to join you.

Although you feel divided by the two groups, not eager to leave your two teammates in the group behind, you are driven to stay within your comfort zone and go where most of the group is headed. You know neither group has solid data or information to guide them. In fact, no one knows where the hell they are going. This fact is scary for you, but what's really frightening is that you will leave the rest of the team behind if you stay with the remaining two from the group. More important, you will challenge the majority of the group, which in your opinion does not seem to be logical, comfortable, or wise. So, you go west with the majority of the group, leaving the two members behind. Imagine now how you feel as you head west. If this is the choice that sounds like one you would make, you may be a "cog in the wheel," unwilling to go against the majority, regardless of whether the group's choice is based on fact.

1.10 What Would a Leader Do?

The one thing the skids on rails, stick in the mud, and cog in the wheel types have in common is the lack of leadership qualities these behaviors exhibit. A leader aspires to engage others, build trust, and work to bring people together. In our story, a leader would exhibit behaviors and take action to:

1. Seek to understand the position of each group—identify and recognize individual fears and what drives them.
2. Help individuals confront and address their fears, together.
3. Rally everyone in the group around a single clear mission—getting to civilization in the safest, most timely way possible.
4. Develop the strategy for action with the group—inspire and empower the group to work together to determine the best approach to accomplish the mission.
5. Model compassion and support others to show care and compassion to each other, sharing water, warm clothing, food, and whatever can help each individual in the group.

In the end, it doesn't matter whether the group decides to stay the night where they are, continue west, or go east. The mark of a leader does not lie in what decision was made. The mark of a leader is defined by how the decision was made and the way the group makes decisions together. A group's ability to stay together, remain united, and support each other through the end of the journey is truly the legacy of a leader. Now, take a moment to think about your experiences as a project manager. What has been your journey? What is the legacy you are creating through your actions and behaviors?

Throughout this chapter we examined some lessons in history, looked at how the forces of the universe conspire to change the factors in the Iron Triangle, and established that nothing you do on your own can improve or change a future outcome that depends on the behavior and actions of other people. Finally, we demonstrated the need to inspire change in the behaviors of others. We looked at the journey through change and examined the need to embrace change and model behaviors which inspire and align others to collaborate to evaluate the facts and move forward together. We established that the project management journey is based on your ability as a project manager and as a leader, to actively and courageously engage in self-discovery and reflection so you can examine your own fears, your capacity for change, and identify leadership qualities to choose a course of action which inspires trust, alignment of the group, and a belief in a collaborative vision to build stakeholder satisfaction and success. Remember, this journey is not perfect. A journey that goes to somewhere you have never been before is not perfect. Just like the project that strives to create something unique that has not been attempted before, your journey will be plagued by challenges and uncertainties. The key to your success lies in your ability to commit to the journey. See success and commit to earning it. It will take hard work, and that is not even the tough part. The tough part is that this journey never ends. The ability to lead and embrace change is not earned in a single effort or span of time. It is earned through your own discoveries, successes, and failures each day, each week, each year, one step at a time. You are never "done" while you are on this journey. Your measure of success lies in your ability to trace the successes and failures of your own performance as a leader, an agent of change, and ultimately as a performance-based project manager. Are you ready?

The true test of character is not how much we know how to do,
but how we behave when we don't know what to do. (Holt, 2018)

Reflect on the lessons in this chapter. We talked about fear as a driver of behavior. Your fear and stakeholder fear can quickly derail your best efforts. Fear can quickly escalate to become your biggest risk on the project. Tell me, how many times has your project team identified stakeholder fear as a "Top 10" risk? If your team has not identified stakeholder fear as a risk, they should. Evaluating your fears and your reactions to those fears takes courage and honesty. Do you have the courage to continue and dive into your fears and the true cost of fear?

Fear of change is one of the top fears that people experience. Fear of change presents a significant challenge on any project because each project represents a journey into change. George Bernard Shaw expressed the challenge of conquering fear to implement change concisely when he said, "Progress is impossible without change, and those who cannot change their minds cannot change anything" (Goodreads, 2018).

To help your team, your organization, and your career to move forward, the first step is to open your mind to new information, approaches, and opportunities, so you can create and inspire new possibilities. The process requires you first to develop your ability to change your approach and behavior in a way that resonates with the individual or group you are trying to reach, while remaining genuine. Progress, leadership, and project management are a perilous adventure through change. Are you ready for the challenge?

Chapter 2

The Firing Squad, Suicide, and Other Hazards of the Profession

© Jim Kangas

Our journey through France in Chapter 1 taught us that leadership in project management takes courage in the face of the unknown. Evaluating your fears and reactions to those fears is the first step to navigating the project journey.

Remember your choices in Chapter 1? What did the journey tell you about yourself? Did you discover that your greatest fear was the fear of failure, loss of control, fear of change, or something we have not touched on yet? Continue to reflect on what you learned in Chapter 1 throughout this chapter to go further on your journey into self-discovery. Be patient as you take an honest look at yourself and the results you can create for your stakeholders. Remember that risk and projects are inseparable partners. On any project journey, you will not find one without the other. Knowing what you are afraid of is critical to developing the ability to adjust your behavior and approach throughout the project journey, because there is a cost to the reactions fear can trigger. Do you know the cost of fear?

2.1 The Day Everything Changed

You learned quite a bit about yourself on the train trip that you and your group survived from Nice to Lille, France. After returning from the train trip, you enter the office on Monday morning with a renewed sense of confidence and energy. You are thinking, "Bring it on. Whatever it is, I'm ready! I'm flexible, adaptable, and agile." Then, your first meeting of the morning starts with your Project Management Office (PMO) colleagues.

The first signal that something is off that day is when the Chief Operations Officer (COO) walks into the room with the PMO Director in tow. The odd thing about this is that Rod, the COO, has never attended a department meeting before, even though Susan, the PMO Director, has reported to him for over a year. Rod's lack of attendance at the PMO meetings has been causing rumors about Rod's interest in the PMO. "Maybe that's why he's here," you tell yourself. "Maybe Rod finally has an interest in project management? It's about time," you think.

As Rod begins to speak, the room is quiet except for the sound of your boss Susan, softly crying. Susan has been the director of the PMO for over five years and you have never seen her react emotionally. You try desperately to make sense of what you are seeing as you hear Rod's words: "The PMO for this company is disbanded as of today." Suddenly, the new confidence you earned as a result of your accomplishments with the group on the French train trip starts to flounder, and the words "Bring it on" that you just spoke to yourself moments ago now spin in your brain. You try to sort out the questions whirling around in your head. Finally, you ask Rod the most important and perhaps the most obvious question: "Why is the company disbanding the PMO?"

In response, Rod explains that disbanding the PMO is a strategic move. This move is designed to remove barriers, so the business can work faster and more efficiently, with the goal of ultimately improving profitability. Looking around the room, you quickly assess that your colleagues are in shock. You find yourself trying to sort out how this could happen when projects are going well. In fact, you reflect, the only projects that have slipped this year were a facilities project and a software project. This means that in the company project portfolio, 10 out of the 12 projects were running well within budget, scope, and schedule. This success ratio far exceeds the average rate of 10 failed projects for every 12 projects. The questions begin to flood your mind: How could the company decide to "disband" project management when performance exceeds the average? How can it possibly be more efficient to dissolve the PMO now, when five of the projects are just three months from go-live? Dissolving the PMO, the sole support structure for project, program, and portfolio efficiency, now would literally put millions of dollars in project value at risk. These thoughts cloud your mind. Still, no one speaks a word. The silence is very unusual because the organization has always encouraged open dialogue and collaboration while adopting the mantra, "None of us are as smart as all of us" as a cultural cornerstone. Throughout the organization's history, leaders have relied on feedback from the rank and file to identify risks and brainstorm solutions that keep the organization on the cutting edge of the competition while identifying and reducing risks. Clearly, that cultural cornerstone is crumbling today.

When Rod finally ends the meeting, you conclude that two things are certain. As of this moment, your boss and the PMO as you knew it are gone, and tomorrow you meet with the director of your new department. Suddenly, you are thinking your recent adventure in France was not so much of a challenge after all. You find yourself wishing you could be back in that remote area of France instead of at the office today. As a matter fact, you tell yourself, anywhere would be better than here today.

Even though France was a challenge, that challenge does not compare to the events of today because today threatens everything you know: the security of your family, the lifestyle you have become accustomed to, the job you love, and your career—hell, maybe even your marriage will suffer because of the changes that occurred today. You think to yourself, the effects of the changes that were made today will certainly not boost your immune system or contribute to your good health. In a few short minutes you quickly determine this is the worst kind of change you and your colleagues could experience. This is the kind of change that is imposed, poorly communicated, and instantly clouds the future for anyone affected by the change. Yes, you tell yourself, this change is the kind of change that threatens survival of a career and tests character. The changes you experienced today make the ordeal you and the team experienced on the train trip through France feel like nothing more than a carnival ride. You think,

"The events we just experienced feel as if we have been pushed out of a plane without a parachute. After today, nothing will be certain."

When Rob concludes the meeting and leaves the conference room, Susan warns before she follows Rob out of the room, "You'd better brush up your résumés." Although you hear Susan, you get up and walk away, hoping that this is all just a bad dream and you will wake up in the world you knew before, a world where everything stayed the same.

When you get back to the project room, the team members in the room turn to you and ask, "How did it go?"

"What?" you respond.

"You know, your meeting with Rod, how did it go?" the team asks eagerly.

Quite surprised, you say, "Wait, how did you guys know about that meeting? I didn't even know about it."

"Oh, we heard about it last week," said the team, almost in unison. "The Chief Information Officer (CIO) sent an email to all IT department employees to say that changes were coming soon for the PMO here."

"Good Lord, how does that happen? How is it that my team and half the company know that my own department will be dissolved, and I didn't have a clue?" you ask, amazed. "Did this email indicate what might happen to the employees in the PMO department, myself included?" you ask.

"Well, it sounded like they will move some of the project managers to different areas of the business, but that was pretty vague. We thought you might have learned something more from your meeting with Rod this morning," the team exclaimed.

"Well, I didn't," you chirp. Then you continue, "Frankly, it stinks that you guys know more about this than I do."

Sensing that you are upset, the team turns back to their laptops and stares back at their screens. As the day painfully drags on, you try your best to stay focused and productive, but you soon realize this is hopeless. Then, just before 4 o'clock, an invite pops up in your Outlook calendar. It's from Sam Drucker, the CIO, for a meeting tomorrow at 8 a.m. "Oh great," you think to yourself, "here it comes, the unemployment line. I can't live on unemployment with my bills. This could be the beginning of the end." You immediately bury your thoughts and fears, trying to control your emotions. Then you accept the invite, close your laptop, pack your things, and go home for the day.

That night you are haunted by what-if scenarios that seem infinite in number. The thoughts haunt you and flood your mind to create a river of emotion ranging from shock, to fear, and finally anger. You think to yourself, "How can they treat you and your colleagues this way? Are those executives so far removed from reality that they are unable to appreciate the hundreds of hours you labor so the company can actually complete some of the screwball projects that are assigned by leaders who value the political outcome of the project more than the

ROI? Maybe they have no idea that 77 percent of the organizations that utilize a PMO are able to align projects to a strategic vision and clear business goals to help them achieve the project objectives? The company executives must not have read the PM Solutions Research (2014) 'State of the PMO' report, which also clearly demonstrated there was a 27 percent decrease in project failure for those organizations that had an established PMO. What other information do these executives need to prove the importance of a PMO to the organization?"

The thoughts and emotions continue swirling around in your head until the sun's rays begin to break through the darkness in your bedroom window. You slowly squint and allow your eyes to receive the light of day. After a heavy sigh, you say to yourself, "Here we go, let's get this over with."

Day 2

Shortly after you arrive at the office, you find yourself standing outside Sam's office awaiting the 30-minute scheduled meeting. You've left your personal affects in the car and have your laptop with you in preparation for that dreaded march out the door. Sam looks up, calls you into his office, and asks you to have a seat. Then he asks you, "What projects are you working on these days?"

This question seems odd because your projects, along with all the PMO projects, are clearly outlined on the portfolio dashboard, which is designed to provide at a glance the project status and performance measures for schedule, budget, and resource utilization. Each C-Suite executive receives the link to the dashboard and has access to it on the company's intranet site. Knowing how busy the company executives are, you conclude that Sam probably never read the dashboard and has no idea of the status of any of the projects, let alone yours. You obediently answer Sam's questions without further comment. "Hmm, what else doesn't this guy know about you or the projects you are working?" you wonder.

Then the pit in your stomach quickly grows larger. Sam announces that you will be reporting to the enterprise architect. You and another project manager will be retained by IT, another two project managers will be retained by Rod and placed in functional departments throughout the organization, and the rest will be "released." Sam looks at you and, without pausing, says, "You will meet with Aaron, the enterprise architect, as of tomorrow. That is all. You can rejoin your project team now and continue your work. Let's just get that project reeled in by the go-live date. No slip-ups. Got it?"

You are too dumbfounded to respond. All of a sudden, it feels like the whole damn group of project managers was just put in front of the firing squad and you were left on the sidelines to watch as they were shot dead! You shake off your disbelief and nod to Sam obediently. As you get up to leave the room you think, "Wait, there's 15 minutes left in this 30-minute meeting. This meeting is not over. Stay, Terry. This is your chance to say something, Terry. Say anything," you tell yourself. Then your inner voice starts screaming, "No, that's not all, damn it! I'm not leaving. I have a lot of questions that need answers." But the voice never speaks out loud. You soon realize that's because your voice has been effectively silenced by the fear that you too could end up in front of the firing squad. Moments later you pick up your things, leave Sam's office, and head back to the team. As you walk into the room, a couple of team members approach you and, without a word, give you a hug.

The trust and bond you have created with the group during the last eight months is unparalleled compared to any other team on which you have served. Sure, it was tough pulling the group together in the beginning, and it took plenty of precious time to come to terms defining the method of agile you would use to support the project processes, the team roles, gaining agreement on decision-making processes, getting buy-in and commitment on the tools, processes, and even the behaviors you and the team would agree to use on the project. Today, more than any other day, you are thankful you made the investment with the team. You take a breath, pull yourself together, and get settled in at your desk as you prepare to lead the stand-up meeting which is scheduled to occur in just five short minutes.

The project team has honed these stand-ups well over the last few months. Usually, all 10 team members report on what they did yesterday, what they will do today, and state barriers or concerns, which typically signal a follow-up meeting later that day, in about 10 minutes or less. Today, the entire team is co-located in the project room, so you decide to add your update from recent events. You inform the team of the events that have occurred to date, the elimination of the PMO, your new reporting structure, and your commitment to complete this project with the team. As the silence settles in the room, you see a wave of fear pass over the faces of the team. You know the most important thing to do in times of uncertainty like this is to get the team re-focused and remind them of the obstacles they have overcome together in the past. Past success can be used as proof that the team has the capacity to overcome a future challenge. You remember that fear can be an instant de-motivator, or it can be a driver, but it rarely earns sustainable positive results in teams. Fear is a powerful weapon, and whether intentional or not, it can swiftly change the direction of almost any endeavor. The objective in this moment is to keep uncertainty and fear in the background and help the team focus on opportunity and their ability to succeed, so you state confidently, "In all this chaos around here, there is one thing I know for certain, we are a group of the most talented, focused, and committed professionals I've been privileged to work with here at Clover. I'm dedicated to completing the journey with you and bringing this project home. Are you with me?"

The group cheers a resounding "Yes"!

"Ok, then let's get focused and do it! I promise to share any updates as soon as I receive them, you can count on that." At that moment, you see the fear on the faces around you dissipate and, as a bonus, you get a few smiles out of the group. Then, you and the rest of the team turn back to your laptops.

After a brief time, Cindy, the business manager on the software development project you are leading, asks if she can speak to you for a minute. "Oh, boy, more good news," you think. "Sure, Cindy," you respond.

As she closes the door to the drop-in conference room you step into, she says, "Terry, I am so impressed with how you led the group in there. I know it is a tough time for you and your project management colleagues right now, but we need to keep this team together. Your positive approach really helps to keep us focused and on track. As you know, we've got 90 days to wrap up this project for go-live or there will be grave consequences for the business."

"Absolutely," you agree. "Thanks, Cindy, for sharing your thoughts and giving me a few positive vibes. I really needed that today."

"No problem," says Cindy. "You are doing a wonderful job and I appreciate it."

"Thanks, I only wish the guys at the top understood the positive returns project managers and project teams are earning for this company," you sigh.

Before the day ends, the invite from Aaron, your new boss and the enterprise architect at Clover, pops up in your Outlook. You accept the invite, close your laptop, and head home dreading the prospect of another day in what has quickly become the Twilight Zone.

As you toss and turn that night, an uneasy feeling comes over you. It's a familiar feeling, and you find yourself searching to place it with past events. Then it hits you. You think, this is how I felt 10 years ago when I worked for Alfa Retail Stores. Why am I feeling the same way now, after all this time? You decide to send an email off to Marge, your old colleague from Alfa, asking her to meet you for dinner later in the week. Marge was a close colleague during those years at Alfa. You are sure she would have perspective on how the events of the past can help you learn to navigate this new uncertain future you now face. Marge was good at helping others examine their own behaviors to learn, grow, and, most important, she was good at helping others avoid repeating past mistakes. It is time to meet with Marge, you think. Yes, it is time. Then, in what seems the next moment, the sun casts its unwelcome light on the room as you rise to face another day without sleep.

Day 3

Before you leave the house, an invite pops up on your phone: "Conference for meeting with CIO, COO, and HR, 8 a.m."

"Ok," you say to yourself, "just stay calm and keep quiet in the meeting. Everything will be fine."

You arrive at the Madison office to find Aaron, your new boss, and two of your colleagues already sitting in the room. On the video prompter at the front of the room, a projection starts to load showing the rest of the group, who are located at headquarters in Chicago, Illinois. As the prompter comes up, you see two project managers, the CIO, Sam, Rod the COO, and two other managers from the business sitting at the conference table. You sit near Robin, one of the younger project managers, who is responsible for construction projects. One of those projects is way behind schedule and over budget. You also know these overruns are due primarily to the fact that no formal survey was done on one of the manufacturing plants that is being remodeled. The lack of a survey created knowledge gaps regarding the structure of the plant and what it might take to complete the work needed to get the plant running efficiently. The estimates were simply a best guess. As the work began, new requirements were continuously discovered, which ultimately caused cost and schedule estimates to be way off, resulting in schedule and cost overruns. Robin does an excellent job. She is known as a project manager who cares about the organization, her team, and her colleagues. She grew up in the organization, starting as a project administrative assistant to the Project Management Office. Today, she is a valuable asset to the organization who has earned her Project Management Professional (PMP) certification. You respect her. Sitting across from you is Aaron, your new boss. An affable guy, Aaron has been a contract employee at Clover for over a year. Then, two weeks ago he was hired on as a full-time employee. Over the past year, you have worked with him and think of him as primarily a numbers guy who is great with detail and data. Truly one of those left-brain types, he struggles with the emotional side of things, suffering from a low level of emotional intelligence and poor relationship management skills. The meeting begins when Rod starts to speak. You are wondering, "Where are the rest of my colleagues from the PMO?" There were 15 employees in the PMO before it was "dissolved." Then you hear Rod say, "Susan, the Director of the PMO and the rest of the employees in your department have decided to part ways with Clover." The faces of your remaining colleagues are frozen in fear. You sit quietly and listen as anger and frustration wells up inside you. Somewhere in the babble, you realize that Sam and Rod are having a conversation about the rest of the project managers in the room as if they were not even present.

"Rod, do you think we should consider changing the titles of the project managers?" Sam says.

"No, Sam, I don't think we need to do that. Let's just have them report to their new managers in the various functional areas of the business for which they will be assigned. Managers, what do you think?"

You are numb. You look at Robin and can see the disbelief on her face. Then you hear the next part, which is even worse. Rod and Sam start to talk about a project that you worked on for over a year as an example of a successful project the business did without project management support of any kind. Sam states, "Yes, Project Eaze is an excellent example of why it makes good sense to empower the organization to run their own projects. Frankly, we don't need the overhead of a project manager in addition to another manager in the business to guide the team. We can trust the business manager to do it all. We anticipate this means full-time project managers will be used just on our biggest, riskiest projects that involve a large capital investment and take 12 months or longer to complete. This move will help Clover work smarter, finish projects earlier, and spend less money doing it. It's brilliant."

The misinformation in their conversation reeks and begins to make your brain dizzy with confusion. Everyone else in the room knows that Project Eaze was a total disaster. What are they talking about, you wonder? The true story is well known throughout the organization. After a year of floundering with the vendor, the business rang what you would affectionately call the "project management hot line." It was a fast track for the business to get support from the PMO when they tried to manage a project that was too big, too risky, or too complex. Finally, when the project basically ran off the rails as a result of a lack of project management, the project hot line would be used to get the derailed project back on track.

Throughout your career, this type of scenario has been the trigger for your assignment to the project. Two years ago at Clover, a call from an executive promptly triggered your assignment to the project. The goal was to get the derailed project back on track, minimizing risk and loss to the organization in the process. Today, you clearly remember that it took a team of security, data, and IT experts countless hours and double duty to keep the projects they were already assigned to on track while completing the rescue of the derailed Eaze project. Everyone in the company knew this story. Yet, at that moment, no one corrected the erroneous spewing of information that was filling up the meeting room. Even you remained silent. How could you speak? Half of the department is now gone. Who knows who will be in front of the firing squad next?

Then Rod addresses the project managers, saying, "I want to reassure you this was a strategic business decision. This is not a punishment and there is no complaint about the value you all produce for Clover here."

Your mouth does not move, but your mind screams, "What the hell? Well, there may not be stated complaints, but clearly the PMO did not produce value for this organization because it's been removed from Clover in one swift executive move."

Even as this was happening, you knew the value of the PMO was clear to everyone else you worked with in the organization. The PMO was aligned with the organizational strategy, processes were defined, communication was transparent, and it was a structure that had built a firm foundation over the last eight years based on PMI methodology and best practice. Sure, the PMO had to play bad cop occasionally when the business tried to do something that was just, well, stupid. "How could anyone not value that?" you ask yourself. Then a haunting thought settles in: "What do the executives at Clover value? Do we really know their perspective? Has anyone even asked?" Do they have pain points from their experiences at former organizations that we should have identified?"

After a hard gulp, you tune back into the conversation in the conference room and realize the meeting is ending. On your way out of the conference room, Aaron stops you and asks if you would like to stay and debrief with him. "No, I don't think I'm able to stay, Aaron. I'm not feeling well. I think I'll just go back to the project room," you quip.

Once outside the meeting room, you stop in the hall, pull Robin aside, and ask, "Are you OK?"

Robin looks at you with tears in her eyes and asks, "What happened in there? They talked about us like we were not even in the room, like we were nothing. How can I not matter when Clover has trusted me to lead a $30 million plant remodeling project that will go live next month? What's going on, Terry?"

"I don't know, Robin. It feels like the whole world is upside down. Hang in there, that's all we can do, I guess," you say empathetically.

Frustrated, confused, and deflated, you muster up the strength to go back to your team and generate positive energy to keep the group moving forward. As the workday comes to a close, you cannot wait to leave the office and the day behind.

That night you toss and turn. As the day's events rumble in your mind, you are not able to overcome your frustration over the misinformation that was spewed in today's meeting about that example of a "successful business project." Finally, you give in to yourself, pop open your laptop, and begin an email to Sam, Rod, and Aaron to set the story straight on what really happened on the Clover Eaze project.

Day 4

The next day, when you arrive at the office, you find an invite for a two-hour meeting from Aaron for later that afternoon. "Oh shit, is this about the email I sent last night? Damn it, why can't I just let things be, put my head down, and stay quiet?" you think to yourself.

At noon you, Robin, and a couple of the business analysts meet in the café for lunch. Robin says, "Yesterday was awful. I just went home and cried myself to sleep."

"I'm sorry about that, Robin, hang in there, we will get through this," you say.

Robin looks up and asks, "Will we, Terry? More than half the department was fired and I'm doing everything I can to keep this remodeling project from failing. I don't know if I can do it. The contractor is asking for the building survey. It should have been done before we bought the old plant when Dan went out there to evaluate the ROI well before the project was even approved. Now, I find out the executive management team approved the purchase with a half-baked ROI and no formal survey of the existing facility. What it boils down to is the budget, scope, and schedule for this project was based on Dan's impression of ROI and what it would take to get the 100-year-old plant into the 21st century. It's a shit show! How can I hold the line when the budget and manpower it will take to get the plant up and running is now at double Dan's original estimate?"

"You've got a lot on your plate Robin, that's for sure. I know it's a challenge and I'm here to support you any way I can."

"Thanks, Terry. I really appreciate that," Robin replies.

You say, "Robin, what do you think about pulling the remaining project managers back together for lunch tomorrow? It might be good to reconnect."

"That's a great idea, I'll send the invite," Robin offers.

The group breaks away from the table, and you walk back upstairs to the project room to prepare for the meeting with Aaron and another fun afternoon at what has quickly become a different organization than the one you knew and believed in just a few days before.

When you walk into the conference room five minutes early, Aaron is already seated with his head in his laptop. He looks up and asks, "How are you doing?"

"As well as I can, I guess, considering what's happened in the last week."

Aaron replies empathetically, "Yes, I'm sure these events have rocked your world, but I'm here to assure you that you are valued, and Clover needs you now more than ever. We are long on product and sales and we are well below last year's numbers. The truth is, if we don't find a way to turn things around fast, we might not be able to pay our bills at the end of the year."

"Wow, that does sound bleak," you reply. "If project management is performing as well as Rod explained in yesterday's meeting, what's all this got to do with me? If this meeting is about the email I sent yesterday, I'm sorry if I was too outspoken. I just couldn't let the misinformation go on about the Eaze project. I thought it was important that you and the other executives really understand what happened with that project, which they were using as a great example of a do-it-yourself business project."

"I get that," says Aaron. "I checked with the IT security team and it sounds like that thing was really a mess. From what I heard, it was a good thing you were assigned so you could pull in the right resources to mitigate the security risks. I guess the business had no idea that the cloud application they purchased had insufficient encryption for the information they wanted to manage through the application?"

"That's the thing, Aaron, that scares me about the business doing their own projects. Will they know how to identify gaps, assess risks, and make informed decisions without access to the right resources before they start implementing a new technology? The business needs to access experts who can dig in, elicit requirements objectively, complete an options analysis, provide resource and schedule estimates, and so much more before they implement a solution in the organization. I care about Clover, my job, and the people who work here too. I will do just about anything to speed up the return of value to the business from projects, but I just don't get how we are going to accomplish that without putting Clover at risk if we do projects on a large scale without a PMO," you plead.

Aaron replies, "We are working on that answer as we speak. In fact, Sam is determined to announce the new process for project requests, approvals, and implementations in three weeks."

"Wow, that's amazing. So, you guys have all this worked out? What professional framework did you choose? Are we implementing a scaled agile framework, or a hybrid version? Say, did you see the information I sent Rod and Sam on that awhile back?" you ask enthusiastically.

"Yes, Terry, I did, and no, we are not looking at project management frameworks."

Dumbfounded, you blink and say, "Well, can I ask what models of portfolio project management you are looking at if you are not looking at models from project management?"

"None. This is a new world, Terry. You will need to adapt. We should probably get one thing straight right now. Sam doesn't like the word 'portfolio,' so we will now be using the word 'roadmap' instead of 'portfolio' when referring to a portfolio," Aaron blurts.

Not knowing how to process this statement, you reply sheepishly, "But roadmap and portfolio mean two different things. How can we use roadmap when it has a different definition, I mean it's already a word with meaning—right?"

"This is just something the CIO is set on, so we are going to do it," Aaron states crisply.

"Ok, Aaron, sure. So, does this mean we will be retraining vendors and consultants on the new meaning we've assigned to these industry-wide terms every time they engage with us? Is that our vision of streamlining project management and increasing productivity and throughput?"

Aaron sits quietly across the table writing something furiously in his tablet. Then he looks up and glares at you. "Thank you, Terry, that's all for now."

Taking this as your cue to exit, you go back to the team room and stare blankly at your laptop screen, trying to figure out how you ended up in this nightmare.

Shortly afterward, a brief email communication comes across from Rod for you and your remaining project management colleagues. The email states simply, "Until further notice, the project dashboard will be discontinued. Status reports will not be collected or reviewed with teams. Each business unit working with one of the remaining project managers can determine the type of information they need from the project team. This will reduce meeting times and paperwork while increasing efficiencies on projects."

"What?" You yell out loud before you can stop yourself.

The team turns to you and asks, "Are you OK, Terry?"

"Um, yes, I'm fine, sorry, guys."

You pack up your laptop and head home for the day. The world is upside down, the pit in your stomach has grown to the size of a black hole in a galaxy not so far away. Then you remember that tonight you are meeting Marge for dinner. Just in time, you think to yourself, just in time.

It is great to see Marge. It has been five years since you saw her last. She looks great. "What are you doing these days, Marge?"

"Oh, I'm doing consulting. Actually, it's more like coaching."

"Really, what inspired you to go down the coaching path?" you ask eagerly.

Marge replies, "It kind of just fell together. A few years ago, I worked for a medical software company. They produced software that controlled medical cancer therapy. The leadership thought they had to treat people with rigid control and fear tactics to make the company more profitable. Managers and supervisors routinely used intimidation tactics to get things done quicker. People were literally afraid of some of the managers there. Retaliation for mistakes was swift and frequently based only on the information gathered from the manager's observations or hearsay. Root-cause investigations were thought to be a waste of time. Soon, people learned to adopt counterproductive team behaviors to survive and remain employed. A project team was behind on their deliverables and the manager said everyone on the team would be fired if the new software was not ready for launch by the following Friday. This date was arbitrary and two months prior to the estimated go-live date for the product. Corners were cut to make the deadline. As a result, the software was never fully tested. Then, like magic, it went live on that Friday as required. The product was promptly installed in a well-known cancer center and immediately used to control radiation doses prescribed for cancer patients. Two months later, five patients died due to receiving radiation doses a hundred times greater than prescribed. Almost immediately after the deaths, an investigation was launched to examine the software. The investigators discovered the root cause of the deaths was a bug in the software that created random variations in prescribed doses. When the project team was interviewed, they said they knew cutting corners on testing was risky but thought the greater risk was to speak up, delay the launch, and lose their jobs.

"On that day, I witnessed the results of fear tactics first-hand. Lives were lost, careers were ruined, and the good people who were a part of that team will always carry the burden of their role in this tragic story. It literally broke people. Some members of that team didn't work again for a couple of years, and some changed career fields. That's when I realized my mission. The primary value of what we do in project management is not just about producing deliverables like working, defect-free software, it's about leadership. It's about leading ourselves and others to a higher standard of care."

"Wow, Margie, that was quite a story. I get the importance of our work and the need to produce quality deliverables, although I'm not sure what you mean by a higher standard of care," you ask.

"Terry, one of our primary objectives as project managers is to motivate people to produce quality deliverables. The way we motivate others is key. We can use fear to gain compliance without regard for the consequences, or we can use empowerment, creating a culture of respect and trust to build a productive, positive environment for optimal team performance. It's a choice. Serving others as a coach, my mission is to help people make the right choice to develop the behaviors and characteristics they need to create a culture that promotes the best collaborative, productive environment possible."

"That's great. I wish you could implant that attitude in our executive team. In just a few months, our culture of collaboration is almost gone, and people are shutting down and withholding communication for fear they will say or do something that will get them tossed out the door. It's awful."

"Terry, you do know you can empower yourself to preserve a collaborative, productive culture, right?"

"Are you kidding, Marge? I'm one person, what can I do?"

"Be a leader, Terry, be a leader. Walk the talk and model the way for others. You can't change anyone else's behavior, so just keep them from changing your positive, productive behaviors. Don't give into your fears, because giving in to your fear may have greater consequences than you can possibly imagine. Remember my story?"

You think about this for a moment, realizing the power of Marge's message. Suddenly you feel hope. Suddenly, you feel empowered and ready to lead.

Day 5

The next morning at the team stand-up, you provide your updates, explaining your new reporting structure and the change to eliminate the project status reports, to which you add, "On our team, we will continue to prepare status reports, monitor budget, timeline, and scope, and manage those factors so we can keep our efforts on course and bring this project to fruition. Without documentation of where we are at, where we are going along the way, and how the project status measures up to our original plan, I have no way to quickly address questions or concerns that might come up from our sponsor or other stakeholders regarding how we are performing compared to plan. Guys, I know it's a pain to measure the time you spend on this project, but I need your help to keep this going so I can prove at a moment's notice the current status of our investment and the return we are delivering to the organization. Does that make sense? Are you with me?"

As you predicted, the team is on the same page and agrees to move forward with the status reports and all the processes you have previously been using to monitor and control scope, schedule, budget, and resources on the agile project.

Finally, noon arrives and it is time for lunch with the other remaining project managers. Only George, one of the other project managers, joins you and Robin for lunch in response to Robin's email. You struggle to fight off the feeling that being a project manager at Clover these days is not something that boosts a person's popularity.

When the three of you sit down for lunch, it takes considerable effort to turn the conversation away from recent events. Then Robin asks, "How is your team doing, Terry?"

You smile and reply enthusiastically, "The team is doing great. I feel so fortunate to work with this group, they really are the best in class." Then you add, "Even so, we've had our struggles. The board threw a few sweeping changes that impacted finance, sales, and customer service departments earlier this quarter. Those changes had to be implemented immediately, which meant I lost access to the product owner for at least two sprints. Without feedback from the product owner, the developer slowed down and decisions were delayed. It set us back about 30 days."

"Ouch," winced Robin. "Was there any way to avoid that?"

You sigh and respond, "None that I know of, unless Clover wants to backfill their management team before they are assigned to a project, and I don't see that happening in this environment."

"Hmm," says Robin. "I wonder if those guys on the top floor have any idea of the impacts that resource constraints and unknown requirements have on the value we are trying to create from these projects. Do we actually tell them any bad news, or just deliver it in a status report?"

You answer, "Those are good questions, Robin. I for one would be happy to deliver bad news or news of any kind to my executive sponsor, if he would only meet with me for five minutes."

Then George interjects, "Well, you know how it goes, Terry, project management doesn't rate an invite to the top floor unless they are about to walk you out the front door!"

"Um, that was pretty insensitive, George," you say. "It sounds like you're bitter."

"Why shouldn't I be?" snips George.

"I agree the situation stinks," you say. "Let me ask you a question. If we were really good about proving the value of the PMO and the value of project management here at Clover, do you think we would be in the same situation we are in today?"

George frowns at this and says, "What more could we do, Terry? We did our reporting, met with our teams, tracked communication, and managed stakeholders. Establishing the value of the PMO and project management isn't our job, that was Susan's job, right?" George is looking around the table with a slightly confused expression.

"Well, you ask, "if we don't establish value every day so it aligns with what our stakeholders think is important, how do we expect to develop champions in the business who understand how project management positively impacts the bottom line?"

Then Robin asks, "What do our stakeholders value?"

You look around the table and realize that none of you has an answer to the question. "Say, did we ever have an audit done on our PMO?" you ask. You see your colleagues' heads shaking "no."

"No audit that we know of," says Robin.

"Maybe if we had asked for one, we would have been able to pinpoint the gaps in our own perceptions," you reflect. "Clearly, we are misaligned with our stakeholders."

George then informs the group, "Dr. Ginger Levin wrote about this in her 2016 article for the PMI® Global Congress, 'The Future of the PMO: Beyond Benefits and Value.' I was just reading it last night."

"It's all too little, too late, I'm afraid," you warn.

"Well, I guess it's time to get back to the grind. See you all tomorrow, I hope," says George.

The conversation with Marge from the night before still echoes in your head, so you take the opportunity to walk the talk as a leader and respond to George, "George, remember, you aren't powerless here. Think about this as an opportunity to show your team leadership. Continue to model positive behaviors, including collaboration and the communication of the good, the bad, and the ugly. This type of action shows our teams we support and encourage behaviors which sustain a productive, innovative culture, not a fear-based one."

George looks at you for a moment, and with a nod and a smile gets up from the lunch table and heads to his project room.

Time passes at the office and things settle down a bit. The project team continues to make measured progress. As the months pass by, things finally start to settle down a little bit at Clover.

2.2 Professional Project Suicide

One morning you walk into the office and head up to the project room. On your way, you see Robin crying in the hall. "Robin, what happened? Are you OK?"

"Oh, Terry, did you hear about Tim?"

"No, you're the first person I've run into this morning. What the hell happened now?" you ask.

Robin whispers, "I guess Tim wasn't formally tracking impacts to the project budget that were caused from the changes the sponsor made to the scope. When they decided to hook another plant into the Enterprise Resource Planning system (ERP), this expanded the project's scope, negatively impacting the budget, and the schedule. You know the ERP project has been dragging on for five years and it's already over budget due to all the changes that have been made. The sponsor said Tim didn't tell him what was going on and now the budget is significantly over. To keep the project going, they are going to have to ask the board for more funds. Tim is upstairs now. Oh Terry, I think they are going to fire him."

"Robin, do you know if Tim has a record of any communication that shared the status of the changes to the project stakeholders? Do you know if Tim even processed a change request?" you ask.

"Well, I'm not sure, but when I talked to Tim after that email went out telling us not to do status reports, Tim said if they don't want communication he wasn't going to provide it."

"Oh, Robin, I hope for Tim's sake that wasn't the case. It's professional suicide not to have a record of a change or status on a project, documenting any approval and the impact the change might have on project scope, schedule, budget, and quality at a minimum. At the very least, Tim should have continued the reporting to keep the sponsors informed on the status of the project budget and warn them about the risk of cost overruns," you reply.

As your conversation with Robin comes to a close, Tim walks past you, jacket in hand and with a Clover security guard at his side. Tim's face is white. It's as if you are watching a ghost rising from a body that has just committed suicide. Tim walks out the door and into the parking lot, and that is it. It's over.

Robin sniffs and walks quickly back to her area. You walk numbly back to the project room wondering who will be next, and how you will keep yourself and your team motivated for another day. After all, who could possibly continue to walk the talk and not succumb to the fear in this environment? It feels like project managers are being eliminated all around you.

2.3 Moving Forward

That night you are not able to push the questions out of your mind. What does management really value? Speed, quality, experience, delivery? No doubt there are lots of options and if you asked your sponsor, you know the answer would probably sound something like this: "I want all the options. Deliver it fast, cheap, and good. Be a servant leader to the team. Drive them no matter what it takes so I get the deliverables I want fast, cheap, and good. Don't give me too much reporting, just give me enough so I know what's going on and why and give the information to me exactly when I need it, not before, and not after. Don't ask me to prioritize because everything is important to this business. You know what to do, so just go do it."

You smile and laugh out loud at this thought. "Sure," you think, "that's simple. I'll just go do it." Finally, you drift off to sleep.

The next day you wake up determined to walk into the office with a positive attitude, courage, and determination. When you arrive at the office you spend the morning with the project team. Everything is going well. Things are solid and on track for delivery. Team members are performing in their roles and are finally able to focus on the work now, rather than on the process, because the consistency of process and cadence on the project has made them comfortable. Over time, they have become confident in the way they work together as a team to simply get more good stuff done.

You know that the business analyst is focused on eliciting requirements from the business and feeding the information to the developers, so the team can consume the information to write the code which creates the functionality in the application that the business needs. The primary role of the product owner is to make nontechnical decisions and feed the project team information on how the business works and why. Any decisions that need to be made from a nontechnical perspective on how the application will look, feel, and function are owned by the product owner, who also serves as the subject-matter expert (SME) from the area of the business the deliverables will serve.

As life goes on at Clover, you continue to recall your conversation with Marge and resolve to accept that while you cannot control how others choose to lead, you can lead the way to a positive, sustainable, and productive work environment, one team at a time, one project at a time. As time marches on, you continue to look in the mirror, reflect on your behavior to redefine your characteristics that support the leader you want to be while committing to change the behaviors that do not support positive, sustainable, productive results. Still, you are haunted by the questions that remain unanswered. What do stakeholders value from project management? Why did leadership at Clover not value the PMO? You know the survival of project management at Clover is dependent on identifying and delivering stakeholder value. But how can you find the answers to what stakeholders value?

2.4 Avoiding the Hazards of the Profession

If you have been a project manager long enough, you have dealt with the reality that project managers are typically placed in the unfortunate position of making the impossible possible. Often, the expectation that project managers and teams complete projects on time, on budget, and within scope are set even before the final charter is signed. In fact, the guarantee that project managers can perform this type of magic is the reason why any business hires a project manager. Isn't it?

This might even be a reasonable measure, if budgets, scope, and timelines never change. Unfortunately, in today's world, there is nothing reasonable about the assumption that scope, timelines, and budgets will remain constant throughout the life cycle of a project. The most reasonable expectation is that the factors of the Iron Triangle (also known as the Triple Constraint) will flex and roll with a fair amount of frequency throughout the project. So why do so many sponsors, investors, and other stakeholders hold on to this fantasy? Here is a better question: Why do we let them?

By spending time, as Terry did, examining our own behavior and the behavior of others, we discover that fear is a powerful tool in manipulating behavior. In fact, fear may even be the most powerful tool in the arsenal of weapons, capable of quickly changing a culture and destroying collaboration almost overnight. Do you now believe that understanding what stakeholders value from project management and even the PMO should be a top priority for every project manager who would like to stay employed? How do you feel about project reporting? Do you think it is important? When it comes to reporting project status, change requests, and impacts to the Iron Triangle on any project, is reporting equally valued whether you are managing projects that are agile, waterfall, or a hybrid between the two? Do you believe project stakeholder communication is important?

Understanding what tools and capabilities you have at your disposal for a project is key to avoiding some of the hazards that Terry and his fellow project managers experienced in our story. Overcoming your own impulse to respond to fear and instead respond strategically to the dynamics of the Iron Triangle while promoting transparency and trust among all stakeholders will empower you to adapt well to project change, sponsor, and stakeholder reactions, while leading the team continuously to improve, perform effectively, and earn the results that meet the value expectations of the business.

Many would testify that the science of project management, which speaks to learning what tools to use on a project and how to use them, is not terribly complicated. However, the art of project management, that is, the ability to understand when to use which tools and processes, and at what level of rigor you should use them to produce the greatest opportunity for success, while also anticipating stakeholder needs, and supporting the team as a servant leader, can be difficult to grasp. Acquiring the skill to balance both the art and science on a project will help you launch your career to greater heights. Remember what you learned about yourself on the train trip in Chapter 1?

You hold the key to your ability to earn success. The journey begins and ends with you. Are you ready to build your skills to manage the human factor in project management?

Chapter 3

Creating the Freedom to Fly High

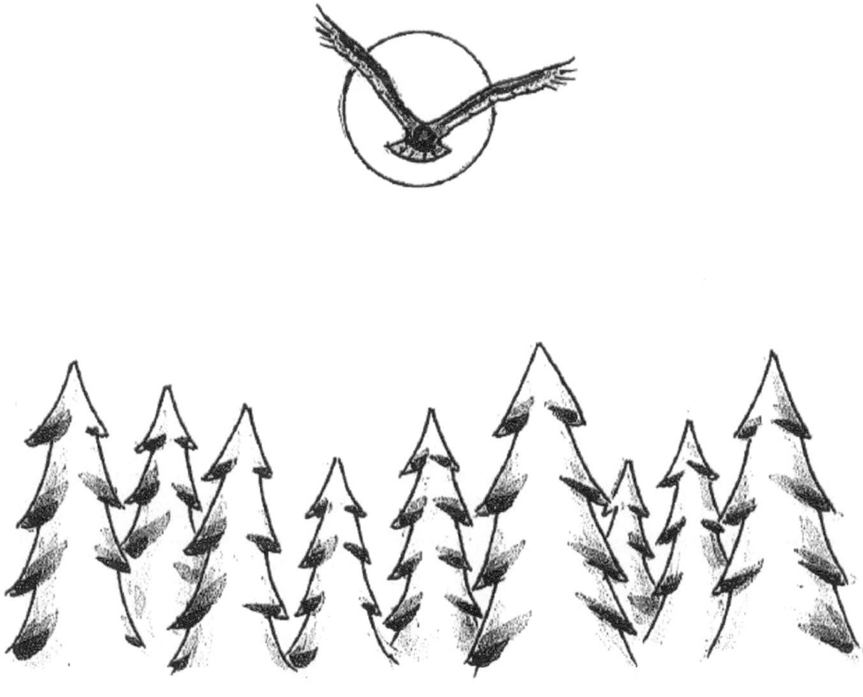

© Jim Kangas

At this point in the book, we have already been on quite an adventure. The resounding theme is based on the belief that each one of us can create a culture of success within projects. We can empower ourselves and others to adopt the behaviors that build a safety net for the team while creating an environment that provides the freedom to innovate and soar to new levels of performance. The journey requires resisting the temptation to adhere to the status quo. Certainly, the journey is riddled with risk, but navigating the risks effectively creates untold opportunities that we cannot even imagine.

Reading a checklist of leadership characteristics and learning about developing behavioral change to build leadership skills in yourself or others is relatively easy. Embracing and adopting those changes so they become part of the fabric of your personality is not so easily achieved.

3.1 The First Step Is the Most Important

The first step is the most important, because it starts and ends with you.

The age-old debate in human psychology of how our personality is developed rages on. Whether you believe that personalities are predetermined in our DNA, or whether you believe they are developed through factors in our environment, there is one common thread: By the time you enter the workforce, your personality is well established.

The *Merriam-Webster Dictionary* (2018) defines personality as a set of behavioral and emotional traits that distinguish an individual. If you have been in the workplace for a while, you have probably taken personality tests such as Myers–Briggs (MBTI®), DiSC®, or something similar. Governments and high-profile employers often require staff to complete personality tests so they can assess individual behavioral and leadership traits. Employers seek to build your personality profile to predict how well you will respond to stress, manage conflict, and accept diversity in the workplace. These areas are just a few of the many key areas of aptitude that are examined. Employers also want to find out what is important to you. They may be seeking to discover your values, ethics, and beliefs, in an attempt to define how you fit in to their culture.

Although personalities are formed early on, personality changes are not only possible but inevitable as we mature. This means you have the power to adjust your behavior to your environment to support your goals. In fact, this process is part of the human tool kit for coping with change. Our understanding of this coping skill can be described most simply by the old proverb, "When in Rome, do as the Romans do." This proverb articulates the advice that we conform, changing old behaviors so we can exhibit new behaviors and customs that are more accepted and revered in our surrounding environment. We do this to become accepted, fit in, and generally get along with others around us.

History is full of lessons that demonstrate the consequence of not adapting to the environment or challenging the status quo. One such lesson is told through the story of the U.S. President Abraham Lincoln.

Abraham Lincoln constantly challenged the status quo of his time. President Lincoln persevered in acting on his conviction to protect the nation and the U.S. government from the rebellion in the South by refusing to pull back federal troops until the surrender of the Confederate Army in 1865. Throughout his presidency, Lincoln set goals that were controversial for many. Lincoln stated the objective and purpose of those goals and stayed the course to see them through, despite the opposition that ultimately ended with his assassination in 1865 by John Wilkes Booth, a Confederate sympathizer. Lincoln's assassination seems like an extreme cost for challenging the status quo (History, 2010).

Lincoln's story begs us to ponder: What is the true cost of challenging others to think, live, and act differently to change the course of history, possibly taking actions that leave an indelible mark on a society? What is the cost versus the return, or the return on investment (ROI) for leading change?

To answer this question, let's examine the impact of other change leaders in history by reflecting on the work of Thomas Edison, Albert Einstein, J. K. Rowling, and even Dr. Seuss (Theodor Geisel). These individuals, in some small or large part, changed the way we think, learn, and live. As a result of the work and lives of these change leaders, the positive impact they've had on society now spans a wide range of cultures and generations. Although all these icons have their own unique story, their individual journeys are linked by one common factor—failure. For Edison, Einstein, and even Dr. Seuss, their failures and ultimately their successes were possible because they each were called to a different path, compelling them to reject the status quo of their time. They consistently challenged commonly accepted standards, and in doing so they did not adhere to the proverb, "Do as the Romans do."

Imagine for a moment what would have happened if Dr. Seuss made his written expressions conform to a professional standard that was more accepted at the time, so publishers were more comfortable with his style? It's true that by tempering his creativity, it's likely Dr. Seuss would have avoided receiving rejections from over 27 publishers. We can imagine that being rejected 27 times certainly provided strong motivation for Dr. Seuss to conform his style to earn success and avoid what seemed to be certain failure. Yet, if he had chosen the path of least resistance to conform his writing style to one more commonly accepted at the time, the world would now be ignorant of the delightful adventures of the *Cat in the Hat, Green Eggs and Ham,* and *Horton Hears a Who!* (Kipman, 2017).

Why do we need to pay attention to the journey of Dr. Seuss? To answer this question, let me ask you, whose work are you more familiar with, Dr. Seuss or Dr. Paul McKee? Both these authors were born within five years of each other. Both were authors of several children's books that were designed to help children read. Both authors dedicated much of their life to writing. In fact, there are many similarities between Dr. McKee and Dr. Seuss. Today, the success of the American author Dr. Seuss, winner of more than seven major awards, including the Pulitzer Prize, in special awards and citations, sets him apart from other authors in his field (The Pulitzer Prizes, 2018). Even though earning these and other awards is impressive, the awards earned by Dr. Seuss are not the reason we are more familiar with him as an author. In fact, I am guessing that you did not even know the exact number of awards Dr. Seuss earned until now. The answer to why we know and love Dr. Seuss lies in the very reason his work was rejected by 27 publishers. Dr. Seuss used an approach that was different. It was that difference that made his approach very effective. Children and adults alike delight in the literary world Dr. Seuss created, because it is a world unlike any other. It is a world that plunges readers into another dimension and carries them on a delightful adventure that teaches us life lessons, gives us a glimpse into another perspective, and tugs at our heart. By pursuing his passion and refusing to adhere to the status quo, Dr. Seuss opened a world of imagination for millions and forever changed the accepted standard for children's books.

The significance of the impacts President Lincoln and Dr. Seuss made on history is quite different. It can also be said that the cost of making that impact is relative to the impact. These stories are a fraction of the examples throughout history that demonstrate the cost of challenging the status quo. Through courage and perseverance, change leaders have the opportunity to make an indelible mark on society. The impact you chose to make as a change leader is relative to your investment. What will be your first step to earning ROI as a change leader?

3.2 The ROI of Change

As a leader in project management, you have an opportunity to influence others and literally change history. If you think that statement is a bit dramatic, recall the story of Ferdinand de Lesseps and the Panama Canal in Chapter 2. Your success or failure as a leader is not a result of the many variables you do not control, or a result of the detail found in your project schedule. It is a result of your ability to have the courage to challenge the status quo and identify when to adopt a standard best practice or tool. Knowing when to innovate is also the key if you want to forge a new frontier that earns success. Of course, keeping your own ethics, professional standards, values, and morals in check along the way is a critical part of the formula.

You can pick up any one of a hundred leadership books and learn that the attributes of courage, curiosity, compassion, commitment, honesty, professional skill, accountability, and the willingness to sacrifice self for something greater are common threads in leadership. The key to your success, regardless of your job title, is to develop these characteristics and use them to provide the optimal environment for project and team performance.

Calculating the return on your actions to drive change can be done to a rough order of magnitude (ROM), at best. To put it simply, you will not have an opportunity to get value out of your efforts unless you fully invest all your efforts. The level of risk you are willing to take on is also a factor. Although the formula is not foolproof and guarantees are not a part of the equation, history demonstrates that big risks present the opportunity for big rewards. It is like making an investment in the stock market. The bigger the risk you take with your investment, the larger the potential reward if the risks do not come to bear. Like any successful investor, the project leader carefully balances the equation between risk and reward, evaluating the potential impact of the risks and the ultimate return. To manage this balance as a project leader you need first to define the level of risk you are willing to accept that will create opportunities for you to make a significant impact on your profession, your career, your employer, and your customers. The greater the risk, the greater is the opportunity for sustainable, meaningful change. In the risk–reward equation, one thing is certain: If there is no investment, there is no return. If there is no risk, there is usually little reward. The level of effort and risk you choose for your investment equation in your career ROI is simply defined by the impact you want to make. As a project leader, you are already driving the development of products and deliverables that, by their very nature, can change the way we think, act, work, live, and learn, creating a sustainable impact on our society.

Project managers are entrusted to lead development of things that create change, so why not lead change in the profession of project management? Why not lead change in the way teams work, act, and think to make your mark in history? The choice is yours and the opportunities to act innumerable. To move forward, you simply have to ask yourself whether you have the courage to challenge the status quo to earn greater rewards. If the answer is yes, keep reading. If the answer is no, perhaps you might want to rethink your choice to serve as a project leader.

3.3 The Leadership Equation

Selling leadership formulas and how-to checklists are big business. If you are looking for a prescriptive list that promises you instant change and success just by reading a list, you have picked up the wrong book. Reading what you need to do is the easy part. I am challenging you to realign your actions for continuous self-improvement that positively builds behaviors in yourself and others which promote project success and satisfied stakeholders. This is the tough stuff you have to embrace if you truly want to lead others. Remember that statement from Chapter 1, "Leadership and project management are synonymous"? If you want to excel as a project manager, you must embrace leading others. Even when you have mastered this, the journey is not over. Keep in mind that there are two parts to the leadership equation. The first is how you behave and walk the talk of leadership. The second is how your behavior is interpreted. Both are critical factors in your formula for success.

3.4 How We Interpret Behaviors

Human psychology tells us there is nothing simple about how we process and interpret what people say and do. A main contributor to our interpretation of someone's words and actions is how we associate what we see, experience, and hear with a former experience, emotion, or thought foremost on our mind (Markman, 2010). The relationship among past, present, and future experiences is an important one to remember as you deal with stakeholders and project variables. For example, people who have been affected by betrayal in the workplace will probably be skeptical and have difficulty believing that others in the workplace are sincere and genuine. Their fear is that they will suffer betrayal again. Despite your best efforts to exhibit behaviors that are intended to promote feelings of trust, you may fail to be seen as a trustworthy colleague. Without knowing someone else's story, it will probably be difficult to understand why your genuine efforts are not accepted. To make the situation more challenging, it's possible that the individual with the fear of betrayal does not consciously recognize the reason why he or she is unable to extend trust. How many times have you heard someone in the workplace say something like, "I can't put my finger on it, I just don't trust that woman." While there is much to be said about the power of human intuition, we need to place equal importance on self-awareness. It is not likely that you will have the ability to make others more self-aware, so make sure you reflect on your own behaviors and interpretations to identify why you decode the words and actions of others the way you do. To succeed, you must accept your responsibility to become self-aware and build positive relationships.

We evaluate behaviors based on our interpretation of present factors. This process is called encoding and decoding. Encoding is the process of putting the information and the messages you want to send into a form that can be interpreted correctly (decoded) at the other end. Decoding is the process the message recipient uses to interpret and understand the meaning of the message, or true intent, similar to our example of the colleague who feared betrayal in the workplace. Consider too that the decoding process relies less on words and more on nonverbal cues and actions. In short, it is not what you say but what you do that provides the input for the process (Meek, 2013).

Remember Sue from Chapter 1? You had a negligible effect on controlling her behavior, so she would leave on time and you both could avoid being late to the party. Through that experience, we learned that although you cannot control Sue's behavior to effect change, you did have control over your behavior. The situation gave you the opportunity to thoughtfully design how you could work with Sue so that you both gained a mutual understanding of each other's needs and the path to fulfill those needs. The next step in the story is to engage in conversation with Sue, build feelings of trust, and move forward in your relationship. Yes, I said, "build feelings of trust." That is because trust is first experienced in us as a feeling. The impulse to trust or not to trust is driven by the amygdala, the emotional and social part of the brain. In the early stages of human existence, it is believed that this process in the brain, which identifies through facial features within the first second or less whether another human is trustworthy, was a survival trigger to flight or fight (Freeman, Stolier, Zachary, and Hehman, 2014).

Today, the same process occurs when managing stakeholder relationships on projects. Short of investing in plastic surgery to change your facial features, the approach I recommend is to focus on how you are decoding information and evaluating behaviors. This will empower you to identify gaps in your own capabilities, assumptions, and behaviors so you can develop sustainable productive relationships. The good news is that we can train ourselves to use cognitive reasoning to overcome our first impulses to trust or not to trust. Your job is to prove yourself trustworthy through actions, words, and nonverbal cues to evolve your team to develop sustainable feelings of trust driven by cognitive reasoning.

3.5 How Easy Is It to Build Trust and Sustainable Relationships?

Ask yourself how much time you invest in thoughtfully designing your behavior to build trust and communication among stakeholders. Maybe you think you do not have time to do relationship building while managing a project. Or maybe you buy into the idea that it is easier to play politics than change your own behavior to produce genuine, sustainable relationships. Is building trust and sustainable relationships really that hard?

To answer this question, let's look at statistics on one of the oldest forms of a relationship: marriage. In the report, "32 Shocking Divorce Statistics," published in 2012 by McKinley Irvin Family Law, the average life expectancy of a marriage in the United States was found to be eight years. Yes, you read that correctly, eight years! Most of us can relate to the struggle of managing, and just maintaining, positive relationships by reflecting on the typical holiday family get-together. From personal experience, you probably also know that if you are striving for a healthy long-term relationship, that requires a whole new level of effort. Naturally, the same challenges we face in sustaining personal relationships also extend to relationships in the workplace. Make no mistake: You have a relationship with your co-workers, your customers, your boss, and your organization. What you make out of those relationships may very well determine your future.

Statistics on Project Management Offices (PMOs) and the general workforce for sustainable long-term relationships in the workplace are even less encouraging. The life span of the average PMO is relatively short, even in high-performing organizations, ending within about six years (PM Solutions Research, 2016). When looking at job longevity for an individual professional, an article published in 2018 by Alison Doyle found the average person changes jobs 10 to 15 times during their career. That equals a job change about every four years or more for someone who is entering the workforce at 20 and exiting at 60. Consider also that, typically, the more stress that's related to the job, the higher the job change rate. These statistics certainly do not inspire confidence in the probability of a long life cycle for a PMO or for the career of a project manager. Here is the good news: You do not have to be part of those statistics.

3.6 What Do You See in the Mirror?

Look at your reflection in the mirror and ask yourself who you were yesterday, who you are today, and who you will be tomorrow. If you are serious about your personal growth and success, those three versions of you will not be the same. Knowing who you are, understanding your capabilities, and identifying the gaps in your capabilities are prerequisites for your growth. While you cannot change the past version of you, the present and future versions of you offer an opportunity to design a different, more capable, successful you. The goal is to create a you that will earn long-term value and achieve greater levels of performance excellence all while building trust-based relationships.

3.7 Developing the Leader in You

Finally, we have arrived at the leadership checklist section of the book, and I can almost hear the sigh of relief. Be warned: This is not a prescribed formula.

Rather, it is a guide that will prompt you to dig deeper to reflect on your own gaps and take steps to close the gaps in your skills, abilities, and behaviors as a leader. Here, I have outlined 10 key areas of focus for you. Each individual journey will be different in your effort to develop the 10 key areas. Engage a buddy, coach, or trusted mentor to help you evaluate the efforts you are making and measure the success you earn through these efforts. Discuss what works, what does not. Find the root cause of failure and reevaluate what different behaviors and actions are needed to address negative results and start over. In practice, you are applying the Deming Wheel, also known as the PDCA: Plan-Do-Check-Act cycle, to commit to self-improvement (Mind Tools, 2018). (See Figure 3.1.)

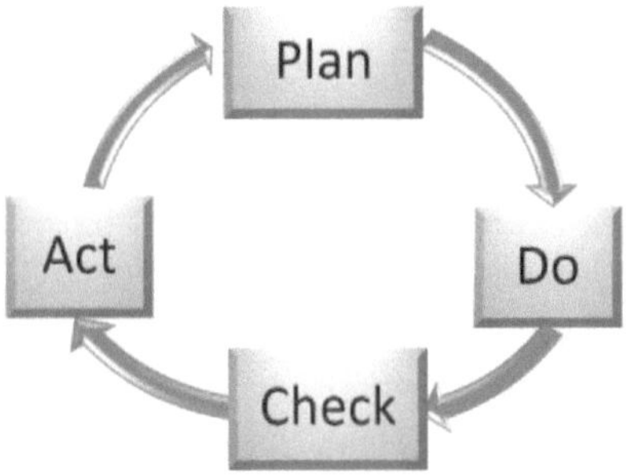

Figure 3.1 The Deming Wheel, Also Known as the Plan-Do-Check-Act Cycle

3.8 The Ten Steps to Leadership

Step 1. Find Your Tiger

This step is all about gathering your courage to challenge the status quo and do something different. To begin, think before you act and measure your actions in terms of the risk and return. Find your passion, engage it to overcome your fears, and drive your commitment to success. Constantly push yourself outside of your comfort zone to improve and stay relevant.

Step 2. Fight Your Demons

As in Step 1, you need to summon the courage to move forward to complete this step. Fighting your demons is more about introspective actions and reflections than about external action-based behaviors. Fighting your demons demands that you look deep inside to identify your trigger points and hot buttons. Examine how you can keep from lighting your fuse or someone else's. Understand what triggers your fears. Identify and define the factors in your environment you need to feel safe and calm.

Identify, know, and understand your fight and flight responses. If avoiding the situation (flight) is typically your response and you are not able to execute that response, how would you behave and what would be the likely reaction to your behavior?

Be thoughtful in situations and design them so you can avoid activating your trigger points, and have a self-management plan in place to mitigate your reaction when a hot button has been pushed.

Step 3. Keep It Real

The idea of being genuine is nothing new on the checklist of leadership behaviors. We also know trust is a key factor in motivating people and inspiring teams. Keeping it real and exhibiting the behaviors outlined in this chapter are building blocks for trust. As a project manager, sponsor, or team member, earning trust is essential to your ability to lead people through change and new experiences. If all this seems a bit obvious to you, think again. A recent study published in *Forbes* magazine found that 82 percent of people do not trust that their boss will tell the truth (Kiisel, 2013). Somehow our ability to act in a way that promotes trust is being compromised in the workplace. Are you actively exhibiting behavior that promotes trust?

Use your tiger to share good, bad, and ugly news. We all know the good news is easy to share. The bad news is not. Sharing bad news openly is what will build trust and confidence in your ability to lead others.

Be honest about your shortcomings and capabilities on the project. Do not cover up your own failures. Do not make excuses. Use failures to evaluate critically and unemotionally how you can use the experience to identify gaps and improve. Use yourself as an example to demonstrate the power of continuous self-reflection and learning. Let others see your human side. Remember, nothing promotes distrust more than the appearance of perfection. Why? Because it does not exist, and we all know it. Put the issues both seen and unseen on the table, so you can deal with the issues collaboratively with stakeholders.

Step 4. Be Last

This is all about your commitment to be instrumental in the ability of others to be successful. Be dedicated to build success for others to achieve the mission. Embrace the responsibility of leadership. Focus on your role to care for others and push yourself to do more for others than you would do for yourself. Do this consistently. Sacrifice self when it offers an opportunity for someone else on the team to succeed. Model the way of care and compassion. This investment will pay huge rewards in team performance, your ability to build sustainable relationships in the workplace, and ultimately your success.

Step 5. Step Up to Step Back

To be an effective leader, lead from behind. This means you give others the limelight, stay humble, take ownership when things go wrong without blaming others, and give credit to others instead of yourself. This behavior shows itself when you seek opportunities for the product owner, the individual who represents the needs of the business and receives the deliverables, products, or services produced by the project. The product owner might walk through product demos with stakeholders as a result of a product created by the project. The product owner may also give presentations to the governance committee and other stakeholders to increase engagement and the ownership of project outcomes. Readily give opportunities to others on the project team to learn and grow in project management knowledge. Trust that each individual is knowledgeable and capable of doing a good job. Trust that when members of the team do not have what they need to succeed, they will reach out if you serve them as coach, mentor, and teacher. Help the team reflect on failure as well as success. Hold the mirror of reflection, not the finger of accusation. When the team reaches a milestone, celebrate their success. When a sponsor congratulates you on a project well done, defer the credit to the team. Remember, you could not have done it without them. Finally, provide an environment where individuals on the project team can own project failures and success. Reward the courage to fail. Do this and you will quickly become the cornerstone of success for your team. Your role is to be the foundation that enables the team to build their skills and success.

Step 6. Commit to the Journey

Commit to the journey of self-improvement, every day.

Are you coachable, approachable, and committed to the team, or are you in this simply for self-gain and promotion? Define how you roll. Are you a solo or team player? Commit to the mission and the team. This means your mindset shifts from a "We have a problem" approach to a "We have an opportunity" approach. See the opportunities, and commit to develop them for the project, the team, and the organization.

The other side of your commitment is perseverance. You will stumble, you will falter, you will fail. Don't give up. Embracing the experience of the journey throughout your project management career will pave the way to greater accomplishments.

Step 7. Become a Trusted Advisor

Teach, mentor, and coach others to gain a deeper understanding of the profession, your own knowledge, and your limitations.

As a teacher, it is not about your story. It is about the learning journey that you facilitate for others on the project. Understand their story and use it to maximize opportunities for success.

As a mentor, coach, and teacher you will be pushed to know your profession and stay relevant. Be a credible source of knowledge and share that knowledge. Know your project. Know your stakeholders. This is more than knowing your project and knowing who your stakeholders are; this is about knowing what your project and stakeholders need to attain success. What is the vision and mission for the project and for your organization? Know it. Commit to it. Live it.

Get the elevator pitch right. Make sure everyone on the team can recite the purpose and value of the project. Why is completing the project important? What problem do the deliverables solve, or what opportunities do they exploit? Why is it important to do the project now? What is the measurable ROI that will be earned when the project is complete?

Step 8. Embrace and Understand Change

Reflect on the train ride we took in Chapter 1, when an uncertain turn of events occurred as the train stopped. Use that exercise to understand how you feel about change. Is it familiar and comfortable for you, or is it foreign and uncomfortable? We all have a different comfort level when it comes to change. Make no mistake: If you are in the workforce, you need to embrace change to stay relevant and to be able to exploit risks and opportunities. In the profession of project management, the ability to recognize, embrace, and adapt quickly to change is essential for the health of your project and for the health of your career. Accept that change is inevitable. If your tendency is to resist change, ignore the call to change, or refuse to embrace change, my advice to you is to stay away from project management as a career choice.

Finally, understand that change for the sake of change is often not productive unless it is driven by a business need. If you are working in an environment that drives change for the sake of change itself, not changing may, in fact, be the change. Try instead to find the root cause of failure, commit to an action plan, and measure the results to readjust and crisply define the changes that are needed to increase success rates.

Step 9. Create a Circle of Safety

Practice empathy, seek to understand, and ask why first. Exhibit these behaviors every day. Wake up daily and answer this question for yourself: "How can I make those around me feel safe today?" Feeling safe in an ever-changing world is a challenge. In today's workforce, jobs, the viability of your position, and even the sustainability of your employer and its customers are uncertain. Your ability to create a circle of safety for those around you and within internal and external customer groups will create the safe haven that everyone is seeking. It will create an environment of understanding, trust, and acceptance.

I am not saying that incompetence, low quality, negativity, or lackluster team performance can be ignored. To the contrary, these factors are exposed when the team feels safe to share failures, ask for help and forgiveness, and seek improvement. When the environment feels safe to the team, the team feels a responsibility to help others succeed and feels comfortable in asking for help when they are falling short of the success required. This environment is a prerequisite to understanding the root cause of performance gaps and to working together to close the gaps. In this circle of safety, you are the mentor, coach, teacher, and leader. You model the way through the behaviors you exhibit every day. To do this, you will need to employ the other nine steps to leadership.

Step 10. Define How You Will be Remembered

This step speaks to the permanence of our actions.

To understand this step better, consider the following example. When you think of the 37th President of the United States, Richard Nixon, what comes to mind? Was your first recollection the Watergate scandal? Or did you first recall that Nixon's administration started the Affirmative Action program, increased funding for the Equal Employment Opportunities Commission (EEOC), and proposed legislation that created the Occupational Safety and Health Administration (OSHA), as well as founding the Environmental Protection Agency (EPA). Maybe you first recalled that the Vietnam War ended during Nixon's administration (The Editors of *Encyclopedia Britannica*, 2018)? If you are like the average American, you probably remember President Nixon for the Watergate scandal and not for the many other accomplishments achieved during the Nixon administration. This example supports research which concludes we remember bad experiences much longer, and those memories are much stronger than our memories of good experiences (Tugend, 2012).

Undoubtedly, you will make mistakes during your career. Remember, though, it is how you handle those mistakes and your ownership of them that can change the stakeholder experience from a negative to a positive experience. While we do not have the power to move through life flawlessly, without failure, we do have the power to admit our failures, learn from them, and overcome them while modeling the behaviors of honesty, trust, humility, ownership, and courage. Through our behaviors, we can inspire others to remember us for how we respond to failure and how we overcome challenges to lead others to earn success.

3.9 John, the Story of a Leader

It was John's first week as a project manager for OnTarget Solutions. Now that he was inside the organization, John was learning that although the company was "best in class" for producing consumer products, the business and technical architecture of the organization left much to be desired. The Enterprise Resource Planning (ERP) software had been so customized that even the slightest change required more code to assure that integration with the 20 other systems remained in sync with the application. To make the environment even more complex, two core business units, finance and production, were considered the primary owners of the software system and jointly shared responsibility for coordinating changes, customizations, and decisions that ultimately determined how the application would be used and maintained. The IT department served the business to ensure that business needs were met while technical platforms and practices adhered to best practice whenever possible.

Although the business units and IT generally worked well together, they each had vastly different approaches, needs, and requirements. This made the process of implementing best practice for the technology while fully meeting the needs of the business an elaborate dance of communication, negotiation, and compromise which threatened to slip project budgets and timelines to more than double the original estimates.

John's former employer had a Project Management Office. Through the PMO, the organization created somewhat of a templated project environment and approach. The processes used were predictable and consistent even though they did not always produce predictable, consistent results. There was no PMO at OnTarget Solutions. Project Managers reported to the Chief Information Officer (CIO) and floated to different business areas to assist with large projects. Project teams were formed to support enterprise projects. There was high competition for scarce resources that were poorly coordinated. The process was inefficient and expensive.

John's first assignment on his new job was to implement an upgrade to the existing ERP application. A team of five had been working on planning for the upgrade for almost two years under the direction of a contract project manager whom John would be replacing. John spent the first week observing the contract project manager, Steve, interact with the team and facilitate sponsor and steering committee meetings. During that time, John identified one of the main problems which contributed to the long planning timeline: No one on the team agreed on anything. There were two executive sponsors, a Chief Financial Officer (CFO) and Chief Operations Officer (COO), who were never on the same page, and the team consisted of a mix of subject-matter experts from finance and production. These stakeholders did not agree on process, methods, or approach. The organization did not require agreement, nor did it impose mandates; instead, it created countless workarounds when collaboration was unsuccessful. The idea was that this would help the group move forward with an initiative even without agreement on a unified approach. Sponsors would openly disagree in project meetings, stalling forward movement, creating confusion, and paralyzing the team, who were keenly aware of the criticality of the upgrade. It appeared that Steve was personally motivated to avoid further conflict and chose repeatedly to tell sponsors everything was fine and the project was on track. These dynamics continued for a year even after the facts revealed a gloomier picture of project health.

After one week, John had observed enough. He was sure he could introduce new tools and processes while modeling new behaviors to improve project performance. John advised the sponsors to release Steve, so he could transition fully into his new role and lead the team. On Steve's last day, John took him out to lunch for a final review of the project. John inquired, "Steve, I see these status reports have indicated all factors are green on the project dashboard. That means you are stating the project is on schedule, on budget, and resources are sufficient to complete the project deliverables as planned, right?"

Steve said, "Uh huh, that's right."

"I did some digging, and it looks like there is no record of changes on the project. Does that sound right?" John asked.

"Yeah, that's right. What are you getting at, John?"

"Well, I was wondering because, according to the charter, the upgrade should have been completed over a year ago, yet the schedule indicates the upgrade won't be complete for 18 more months, and there is no documentation on changes in the scope. Something doesn't jive here, Steve."

"Well that's because the sponsors don't agree on anything. I can't even get through the agenda items in a meeting. Finally, I gave up trying to push the sponsors to agree. A while ago, I tried to indicate the health of the project was at risk of failing in a project status report. For my efforts, I got my head bitten off by the sponsors. After that, I learned my lesson and decided to just tell them what they wanted to hear. I'm warning you, John; these people don't want to hear the truth if it means sharing bad news," Steve stated defiantly.

"Don't you think that's a risky approach, Steve? I mean, won't someone wake up and wonder why the project cost is triple the original cost estimates and double the original time estimates without the benefit of completing a damn thing?" John stated in earnest.

"Hell no, I'm telling you, just keep everything green and you can keep the peace."

Steve stated this so matter-of-factly that John had to ask, "How does any project manager keep their job when they do that?"

Steve raised his eyebrows like John was a spy from the other side and said quietly, "They don't measure anything, so they can't keep track of failure."

John shook his head and thought to himself, "If they can't measure failure, they can't measure success. What have I gotten myself into?"

The next morning John gathered the project team together for a quick meet-and-greet. After introductions, John said, "I'd like each of you to share what you think is the biggest problem on this project."

Immediately he could see a couple of team members stiffen up in their chairs. The body language of some of the team members gave John the clues he was looking for on the dynamics in the team. He was guessing that the team was not accustomed to "laying it on the table" to share the real issues, and the question he just asked pushed them beyond their comfort level. John knew that by asking this simple question, he was showing the team he was not afraid to challenge the status quo and talk about the problems on the project. He was asking the team to trust him enough to share their thoughts.

Almost unanimously the team said, "We don't know what the sponsors want us to do. There is a lot of work to be done and much of it requires our sponsors to make compromises or change the way we have been doing things for over a decade. Change is not popular around here. To complete the upgrade, we will have to make changes to our processes and departments will have to make compromises. So far, we can't even get the different stakeholders to agree on the final scope of process and system changes that will be implemented in the upgrade."

"The accepted scope, describing the work that will be included in the project to complete the desired project outcome, should have been outlined and agreed to in the project charter," John said inquisitively.

Sarah, the production SME stated, "That charter is useless. The sponsors could never agree on scope, so Steve stopped trying to get them to agree and just moved forward without waiting for agreement."

Slightly alarmed, John said, "Without agreed scope in the charter, how do you know what you are supposed to do to move forward on the project? How do you know what the picture of success is for the project from the sponsors' perspective, and how will you know how that success will be measured?"

Tom, the finance SME commented, "I guess we don't."

"Ok team, one last question for this morning, then we will define working agreements."

"What are those?" Carol another production SME asked.

"Working agreements will give us a way to define our expectations of one another and outline the commitments we are willing to make so we can create a solid, trusting, team-led environment."

Looking around the room, John saw the blank faces of the team. Based on the team's reaction, John was guessing they had probably not been through this type of exercise before. John began, "Here goes the question: Tell me, what is the greatest asset the team has right now?"

Debbie, the business process analyst on the team, said confidently, "Well, we are the greatest asset we have, of course." The team chuckled.

Then Debbie said, "Seriously, if we didn't hold it together in those chaotic meetings with our sponsors, we would be in really bad shape."

John replied, "That's good to know Debbie. Tell me, what does 'holding it together' look like in those meetings?"

"The sponsors want us to review in detail what we will be completing in the next month. The minute we walk through the detail, the two sponsors disagree on the work we should be doing, the order, and the changes that will be required to complete the work." Debbie continued, "Disagreement between finance and production on terms, processes, and how we use our ERP has been going on since we first installed the system four years ago."

"Ok, that makes sense. I still need a bit more explanation to understand how the team 'holds it together' during these sponsor disagreements," John asked.

"We just keep quiet, keep our heads down, and wait for the time to run out," Debbie said unapologetically.

"Thanks, Debbie, for sharing that," said John. Then he closed the conversation with, "Does anyone else have anything more to share on this topic? If not, let's spend the next 15 minutes working on team agreements."

During the following few minutes John laid the groundwork for the next part of the team meeting: "I'd like each of you to define behaviors you would like your teammates to exhibit throughout the project. We will create agreements that call out behaviors that we believe will help us work more effectively together. But before we start our list of agreements, let's set a few ground rules for the session."

Next, John began to list the ground rules for the team.

Ground Rule 1. This is a no-judgment zone. Please keep judgments on the ideas and comments that team members want to include in our agreements list to yourself. Remember, no idea is a bad idea in this session.

Ground Rule 2. Show respect and refrain from interrupting or criticizing your teammate.

Ground Rule 3. Remember, the list we are about to create is a living document. This means that as we move forward together, we can add or delete from this list if we think it will help us improve the way we work.

Ground Rule 4. Everyone is equal in this room. That means no one's idea is more important because they have a bigger title or level of responsibility.

Then John continued, "At the end of the 15 minutes, we will review the list of agreements we create together and refine each one until everyone understands how to exhibit behaviors and actions that support the agreements on the list. Any questions?"

John looked around the room and could tell that he had sparked the interest of the team. He thought, "Ok, this is a beginning at least." After a few moments and no questions, John began to work to create the agreements with the team. Fifteen minutes later, the team had created a robust list of agreements. The exercise seemed to create a new energy in the room. The meeting set a productive tone for the day. It was a good first day.

3.10 A New Day

The next morning, John gathered the team around the table in the center of the room. The team had been working in a relatively small space for their number. They were a group of six team members including John. Workstations fit tightly against the wall in the 16 × 20 conference room with no windows. The ventilation was poor, so the team had to open the door in the afternoon to get fresh air. The walls were blank. It was far from an inspiring workspace, John thought. "Today, we are going to review and agree on what tools and processes we want to use to support our work. We can share and discuss tools such as different types of spreadsheets and project scheduling tools, then match them with the best processes that will make our work more transparent and manageable for the outcomes we are trying to produce. Let's get started by listing the tools and processes you have used on this project so far," John said as he looked at the blank faces staring back at him. "For example, I've seen the project schedule and I know Steve completed status reports for the sponsors."

Debbie responded, "Steve thought it was a waste of our time to review the schedule or see the status reports, so we were not really involved in creating or using those tools."

"Hmm, that is an interesting approach. Without seeing the schedule or reviewing a status report, how would you know what was being communicated to the sponsors about the health of the project, schedule, or budget?" John asked curiously.

"Now that I think about it, this was a source of discomfort for me on the project, because I never really knew what Steve was communicating to stakeholders," Sarah stated frankly.

"Thanks, Sarah, for sharing that," John responded. "Well, it appears we have some work to do today. Should we add a team agreement that the team has access to all stakeholder communications?" asked John.

"Wow, that would be great if you would share that with us. John. I know I'd feel more confident if I knew I had the same information, at the same time, as any other project stakeholder, including our sponsors," exclaimed Carol.

"Sure Carol, we're a team. I'm committed to making sure we all feel equally invested and empowered on this project. We each have a role and responsibility to each other as well as to the project outcome. From this point forward, count on me to ensure project status and general communications are shared with each of you before they are shared with stakeholders outside the core team," John answered.

Then John added, "There's one more thing I'd like us to include in our discussion today. Let's make sure when we select a tool or process that we're able to define the value we will receive by using that tool or process. If we can't describe the value we expect to gain, we'll put the tool or process on the whiteboard and return to it later. Agreed?"

Immediately, John saw a sea of enthusiastic nodding heads, indicating the group agreed and was eager to start making the list.

3.11 Developing the Process

John spent the next several days collaborating with the team to gain clarity on the tools and processes the team could use to reach optimum performance. Then, John suggested tools and project management processes that would provide the value and expected outcome the team needed to succeed. John also worked to define who would use the tools and processes, how they would use them, why, and when they would be used. This helped the team get a clear understanding of how they would work moving forward together and why. John knows that defining the what and why *with* the team and not *for* the team are essential components that will help the team understand the purpose and value of the tools and processes used. Because the team chose the tools and processes together with John's guidance, they have ownership of the tools, processes, and results earned. This sense of ownership will prove useful when there is pressure from sponsors and stakeholders outside of the core team to skip process or throw out tools. Each member of the core team will be able to articulate clearly the risks of not using a process or tool and the benefits of using it. Although this collaborative process may take more time, it will save John and the project team countless hours and potential rework in the future.

3.12 Scrum and Kanban

By the end of the second week, John and the team had set up a mix of tools and process, including a poster Kanban board which was displayed in the project room. This tool, and the process associated with it, was new to the team (see Figure 3.2).

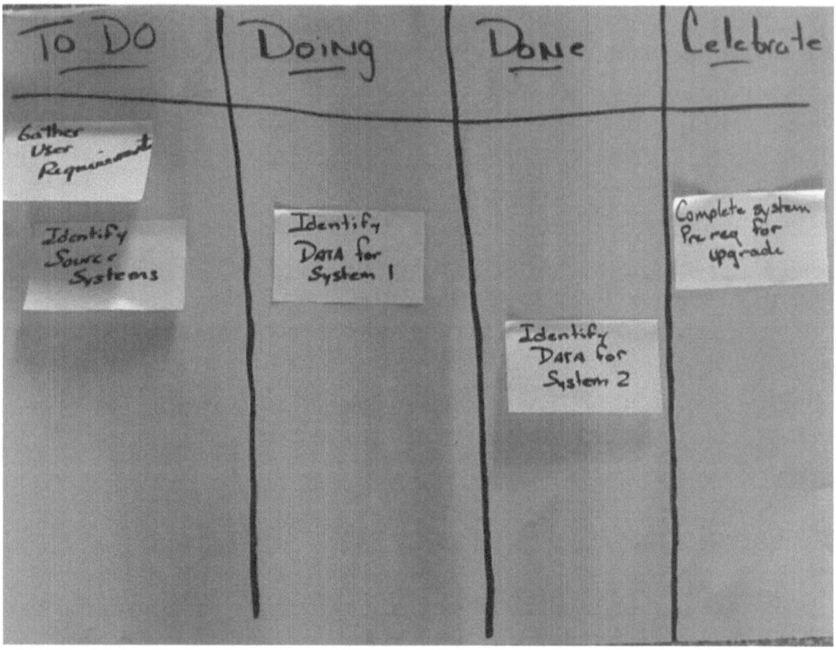

Figure 3.2 The Kanban Board Used by John and the Team

A Kanban board is a straightforward way for the team to visualize the work, what is being done, and what has been completed. In its simplest form, when the status of a task changes from doing to done, the owner of the task moves the card to the "Done" column on the board. This real-time visualization of task completion gives the task owner a sense of accomplishment and the team an up-to-date status of project tasks.

The Kanban tool alone is probably not robust enough for their project. It is a pathway to experience a different, more effective way of visualizing the progression of tasks for the team. A positive experience with this tool will open the door for John to introduce more robust tools that will help the team manage the complexity of the project in a transparent, collaborative fashion throughout the project life cycle.

John also challenged the team to embrace the change in the way the team was working by aligning Scrum framework values with the values of the team. Scrum Framework Values require the team to value people and relationships over process and documentation. After the team was feeling good about using the Kanban board and were familiar with the process, John slowly introduced more processes and tools like the Kanban board which minimize the need for documentation that does not provide immediate value, creating an environment where the team can focus on completing the work and use processes that provide direct value and project results.

Soon, the team had set up two-week time periods in which they planned and completed the work. These were basic sprints. A sprint is defined as a time-boxed event, usually two to three weeks, in which project work is completed. The goal is to complete a deliverable that can be released and used immediately. Every two weeks, the team identified tasks and placed it in "To Do." When a team member was ready to complete a task on the board, he or she moved the task to "Doing"; when the task was complete, it was moved to "Done." Everyone on the team looked forward to reviewing the columns of To Do, Doing, Done, at the end of each sprint. John placed the category of "Celebrate" at the end of the board as a reminder to the team to celebrate what was completed at the end of each sprint. You may be familiar with these concepts if you are a practitioner of the agile project methodology.

What is interesting is that John did not introduce the agile methodology, tools, and practice to the team all at once and demand they adopt them immediately. He introduced concepts through the defined value a tool would provide the team, keeping the process as simple as possible, encouraging practice and refinement of process and rewarding success and adoption of the new tools and processes. Finally, at the end of each sprint, the team discussed what they had learned using the tool. This discussion was part of the sprint retrospectives on the project. Sprint retrospectives were practiced at the end of each sprint throughout the project and gave the team the opportunity to examine what they had learned, and what and how they will do things differently to increase efficiency, and to celebrate their success.

John was careful to stay in the background during these discussions. John did not share his opinion, and he did not presume to be a SME. Instead, John used well-placed questions to get the team to probe deeper, so they would gain a better understanding of their experience and learning. Soon the entire team understood and respected one another's perspective, learning that each added value to the discussion. When the discussions for the sprint were completed, each member had a clear understanding of what changes needed to be made to their process and the value those changes should provide.

3.13 Earning Trusted Advisor Status

As the team began to experience the efficiencies of their new way of working, John was now seen by the team as a trusted advisor. This result took several weeks of dedicated work with the team. It was important for John to work with the team daily, so they could quickly form a bond of trust and respect. There was a lot of work to be done, and strong trusting relationships within the team would be the foundation for their success. By the end of the second month with the team, John had earned a reputation as a trusted advisor within the team. He had created a circle of safety and saw the return in his investment as team members openly shared their thoughts, ideas, suggestions, and mistakes. He helped the team embrace and accept incremental change through the introduction and collaborative application of new tools and processes.

Even with all the progress that John and the team had made, there were still several gaps. John knew the work to close these gaps would require fortitude, self-sacrifice, and courage.

3.14 Aligning Sponsors and Stakeholders

By now, John knew that sponsor understanding and agreement on what the team and the project needed to succeed was nonexistent. It was critical for John close this gap, so the team could receive the necessary support to complete the project successfully. OnTarget Solutions did not understand the role and importance of a product owner to project success, so a product owner for the ERP upgrade had not been identified. Without this, the team did not have access to the direction they needed when they needed it from the business on key decisions. The leadership in the organization felt they were close enough to the work to guide and direct tactical activities on projects. Strategic direction and vision were not defined for projects and largely ignored as a factor. Without a PMO, C-level managers were forced to rearrange priorities in an ever-changing sea of current and new project initiatives, swapping the hottest "priority du jour" with projects already in play, delaying existing activities, and creating a domino effect of delayed schedules, changing requirements, and constrained resources. This resulted in low morale at the company and almost zero returns on project investments. The result was a feeling throughout the organization that project management provided little or no value. That feeling was something John was determined to change at OnTarget Solutions. He had to start with leading the way to change the environment.

John set up the first meeting with the CFO and COO six weeks after he started at OnTarget. This was much later than he would have liked. John had a sense of urgency to move things forward in a positive direction, but that sense was not shared by decision makers. Sponsors demonstrating a lack of urgency to empower project managers to use methods designed to restore project health was a personal hot button for John. John was keenly aware of this. He countered his frustrations by reminding himself of his personal mission to define how he will be remembered by those who work with him. This took John's focus to a higher purpose and helped him take a broader perspective on any situation. John had defined how he would be remembered by stakeholders at the end of his career as an effective coach, student, mentor, teacher, and practitioner—someone who helps others do their best work while constantly evaluating and adjusting tools, behaviors, and processes to ensure that the work produces the desired results, delivering value to the organization and a shared stakeholder experience that creates a bond of trust and respect lasting well after the project is complete.

John also had a feeling that there was a lack of understanding in the organization on the importance of roles and responsibilities to project success. He knew that the key to helping stakeholders learn lies in building their desire to learn. This can be a difficult journey because the process is time-consuming. Through the years, John had observed various approaches used by project managers to deal with this problem. A common approach is the one Steve took early in the project, refraining from challenging the status quo, applying anything new, or calling attention to the issues. Steve felt that by not "rocking the boat" he would avoid conflict and keep his job. John knew that while this approach may help to avoid conflict in the short term, it exacerbates project problems, increasing the perception that project management does not add value or produce results. John was willing to find his tiger to keep it real with sponsors and stakeholders, lay problems on the table, and commit to the journey to define how he will be remembered from this point forward.

John walked into the meeting with the sponsors prepared to explain the true status of the project. Over the next 30 minutes, he walked the sponsors through what he had learned so far, explaining how slipping milestones due to a lack of immediate direction by the business led to budget and schedule overruns, how the cultural environment made it impossible to define time frames in which project deliverables could be produced, approved, and released. John kept the information he presented crisp, factual, and high-level, presenting detail only when sponsors requested it. At the end of the 30 minutes, he asked whether there were any questions on the information presented.

Once John was sure that everyone had the same understanding of the information, he switched gears and asked how important was this project to the organization? Both sponsors quickly confirmed that this project was one of OnTarget's top priorities. Then John asked frankly, "As a top priority, is the organization willing to devote the necessary resources, expertise, and support so the project can return measurable value to the organization in the next six to eight months?"

Jane, the COO, said, "What do you need, John? We are here to help."

That was the opening John needed to introduce change. For the final 30 minutes John worked out a set of agreements with the sponsors that set the frequency of communications including customizing the status report to what they wanted to see and how they wanted to receive the information. John highlighted team and sponsor engagement as a requirement to build trust, clarity, and synergy with the team and outlined actions that would help the sponsors become key champions of the project. When communications, roles, and sponsor expectations were clear to both John and the sponsors, John asked the sponsors, "Can you both describe to me your collective vision of success for this project? From your perspective, describe for me what business functionality and capabilities we should have in the organization as a result of the upgrades to the ERP application when the project is successful. Once I clearly understand your expectations, I'd like to make the appropriate revisions in the project charter, so we have a written record of your expectations and I can be assured we are on the same page. This will help me clear up some of the conflicting goals I found in the original charter, which I noticed was not signed. I just want to make sure I'm working with the team to meet your expectations for this effort."

That's when Mike, the CFO, said, "Charters are a waste of time. We don't sign them and rarely complete them around here."

John was prepared for the response and asked, "When was the last time a project met your expectations?"

Jane said, "I can't remember. This might be why we don't value project management much around here."

That is when John asked for the commitment: "That's something I'd like to change. I'd like to implement some simple practices that will help us realize value from this effort before the deliverables are even produced. Would you give me six months to work with you and the team to prove the value of some of the different practices and tools as we use them to complete the upgrade?"

Mike said, "Maybe it's time to try something new. I think it is clear from the facts you have presented to us, John, that what we have been doing has not given us the results we need."

John wrapped up the meeting with a list of action items and a feeling of confidence that he would be able to make a significant contribution to OnTarget through the discipline of project management.

3.15 Managing Project Governance

A month later, the team was ready for their first steering committee meeting. John had met with the sponsors prior to the meeting to create an agenda and set expectations for the meeting. John had been working diligently to get agreement from the sponsors to include the team in the steering committee meetings. He understood that involving the team in these meetings would allow the experts who were doing the work (the team members) to explain any details of the work. This increased commitment and accountability for the outcome. It would also build trust and communication within the team, because they would hear feedback from the sponsors directly.

During the meeting, John acted as facilitator and was careful to step back from the limelight, handing off opportunities whenever possible for the team to ask questions or provide responses to sponsors. John and the team prepared for the meeting together to create the content and the overall message on project performance and outcomes. This inspired confidence and agreement within the team, which resulted in a united message to sponsors. Prior to the meeting John also sent the agreed-on agenda for the meeting to sponsors and the team based on the key truths that needed to be shared regarding the current status of the project (the good, bad, and the ugly) and any other information the sponsors or the team wanted to discuss. He introduced the use of a "parking lot" before the meeting. A parking lot is a tool that allows a group to "park" questions or discussions that are outside of the meeting agenda. Most important, John was careful to keep the team presentation to 30 minutes or less. The remainder of the meeting was reserved for a question-and-answer exchange between the team and the sponsors. John did not lead the discussion during the meeting. He was careful to empower the new product owner, Susan, to lead the group through the agenda. After his first sponsor meeting, John had convinced the sponsors to jointly select one of their most knowledgeable supervisors to fulfill the role of product owner and serve as a dedicated member of the team. John knew his best chance to inspire the product owner to engage and commit to helping the project succeed was to step back from the limelight, so the product owner could embrace her role and take responsibility with the team for project outcomes.

3.16 Developing Team Performance

Over the next six months, John served the team to empower them to do their best work. In the daily meetings, called "stand-ups," the team took turns stating what they had done yesterday, what they would do today, and finally shared any constraints and concerns that might impede their ability to do their work. The team of seven, including the product owner, fine-tuned this process until the average time of the meeting was 10 minutes. As a regular pattern developed within the team for the processes that supported how they worked together, the team gained efficiencies, spending less time on the process and documentation and more time completing the work on the project. John was also careful to listen for cues that revealed risks or strained relationships within the team. He worked with team members to address concerns, acting as a coach and mentor to help promote trust, communication, and a feeling of safety within the team.

John collaborated with the product owner and business process analyst to evaluate the team's processes and tools. New tools were introduced only if the team agreed to them and understood their value. John was careful to keep the tools simple. He applied the tool or process at the appropriate rigor to match the need with the amount of change the team could embrace at the time the tool was introduced. To do this, John worked with team members to identify the tool and process needed, then evaluated the rate of changes that were impacting the team and other factors in the environment that were causing stress and fear. John knows fear is the main cause for resistance to change, so, in times of high stress, John introduced a simple method to use a more robust tool in order to ease the use of the tool and build the team's confidence. Over time, new functionality and process were introduced as needed to provide increased value. John noticed that this approach earned the team's confidence in the usefulness and value of the new tools.

3.17 Earning Sustainable Value

As time went on for the project team, things did not always go as planned and the results did not always achieve the predicted outcomes. Changes in business priorities, competitive environments, and other factors outside the control of the team required John to reevaluate tools and processes to confirm the level of rigor that was appropriate for the gain the team needed to receive by using the tool or process. John noticed that the extra time taken to prepare with the team for steering committee meetings and communicate a united message was especially beneficial when the message on project progress was not favorable.

As the timeline shrank and pressures increased, John continued his commitment to create a circle of safety within the team. He was always careful to guide team members to focus on the "what" rather than the "who" when discussing slipped timelines and failed processes. This also required John to manage his own anxiety over compressed timelines and missed deadlines.

When the project closed, it was six months later than anticipated due to sponsor indecision and disagreement within the business on final scope. Yet the project was declared a success by the sponsors. The upgrade delivered critical new functionality to the business. That value lasted a few years, until business processes changed, more functionality was needed, and the software vendor released a newer version of the ERP application. The sustainable value that John created through building trust-based relationships on the team lasted much longer. Through these relationships, John built his reputation as a trusted advisor. With each new project and team, John focused on building the team and supporting them to do their best work. Sponsors and teams were engaged, results were delivered, and sustainable value was earned. Over time, John was able to gain acceptance of project management best practice, and soon project management was regarded as a valuable asset throughout the organization.

3.18 Reflect

Take a moment to reflect on John's story. How did he demonstrate "The Ten Steps to Leadership"? What results did John create for the organization, the profession of project management, himself, and the team? Typically, project outcomes are measured in terms of return on investment. There is an equally important value to be considered on projects. This speaks to the value of relationships in the organization and within the project team. To create value, a project manager must find a way to challenge the status quo when behaviors, processes, or tools are not creating positive results. This requires self-sacrifice and can make situations uncomfortable due to fear of change and ultimately fear of failure. Project managers are leaders who are called on to summon the courage to lead the team forward to complete the work which produces a unique product, service, or result for the organization. They are relied on to evaluate project data and human factors. Project managers match the information to tools and processes within their project tool box to minimize project variability, increase productivity, and provide positive returns well after the project is completed.

As a project management professional, you will face the choice almost daily to question the value of a process, a behavior, or a tool, or keep your head down, take the easy path, and question nothing. At those critical moments before you make the choice, I urge you to consider that each project journey is your opportunity to lead the way in building trust and confidence in the value of project management for any organization. Gather your courage and take the opportunity to lead.

Chapter 4

Waterfall, Agile, or Timbuktu—Who Cares?

© Jim Kangas

John's story did not elaborate on the tools and methodologies he was choosing, because his focus was the value those tools produced for the team. In this chapter we are going to dig into John's tool box a bit more. As we do, I challenge you to answer the question that is at the center of what some say is the most heated debate in the profession of project management: What is the most valuable tool in the project management tool box, waterfall, agile, or something else?

A wide array of methodologies, tools, and practices are available to project managers today. In fact, new tools and methodologies are constantly cropping up as project managers continually try to update their comprehensive tool box to keep up with ever-changing business demands. These methodologies are approaches—frameworks, really—that guide the order and high-level activities needed to complete the project.

4.1 What Is Waterfall?

Waterfall is the oldest of the project management methodologies. It is a model of linear project development that has been around in one form or another since the building of the pyramids at the end of the fourth century B.C. The first formal description of the waterfall methodology for project management was published in an article, "Managing the Development of Large Software Systems," by Winston W. Royce, in 1970. By 1985, the U.S. Department of Defense had begun formally using the waterfall model for software development. It was soon widely adopted in business and government sectors throughout the world. Today, the waterfall methodology is still used for software development, construction, and many other types of projects (see Figure 4.1).

Waterfall orders the project work in a specific sequence. The names of the steps, sometimes called phases, vary depending on the project type. For example, the names of the steps will be different for a construction project than for

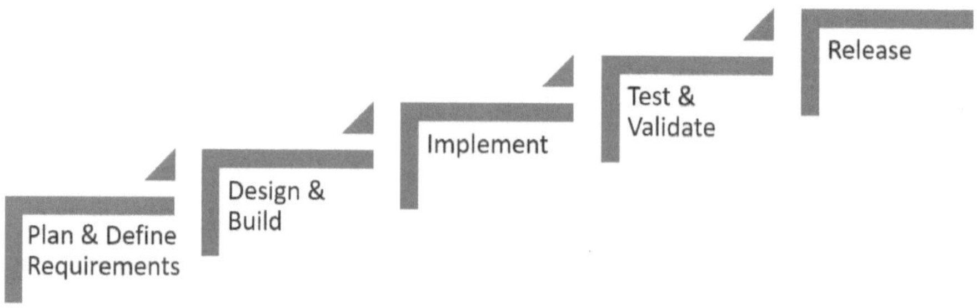

Figure 4.1 A Typical Waterfall Methodology Sequence

a software development project because of the type of work that is needed for the project. Regardless of the project type, the model dictates that planning and design are completed before moving on to the build, test, and implement phases for all project components. This is a very different approach from iterative project development.

4.2 Iterative Development

Iterative development, also known as incremental development, is a process whereby a smaller or individual component of the project is developed to produce a deliverable, which is released before another small part or component of the project is developed. In your business environment, a "release" might mean that the component is tested and released into a working production environment, or it might mean simply that a prototype is approved by the customer. Keep in mind, the best practice for this methodology states that the deliverable should create a piece of working product (software), not necessarily a prototype. The benefit of this approach is that it offers the project team and the customer the opportunity to try out a small piece of the product or result before the entire project is complete. The model follows the Plan-Do-Check-Act (PDCA) cycle, which we discussed in Chapter 3.

4.3 The Agile Methodology

Agile is the third methodology we will examine in this chapter. The framework was formalized through the publication of the Agile Manifesto in 2001 by a group of software developers who saw the need for a method that provides a mechanism to adapt to customer changes in scope and the human learning curve. It stands to reason that there is always a learning curve with projects, because the very nature of projects is to produce something unique. This means the project team has not done something exactly like this before, so it is a safe bet that a learning curve will be a crucial part of the project journey.

The agile methodology or framework is also based on the PDCA cycle of iterative development (see Figure 4.2).

4.4 Comparing the Models

Looking at the agile model, you will undoubtedly notice some of the same steps that exist in the waterfall model. The main difference between the two models is the idea that agile breaks the work into iterations or small components. For

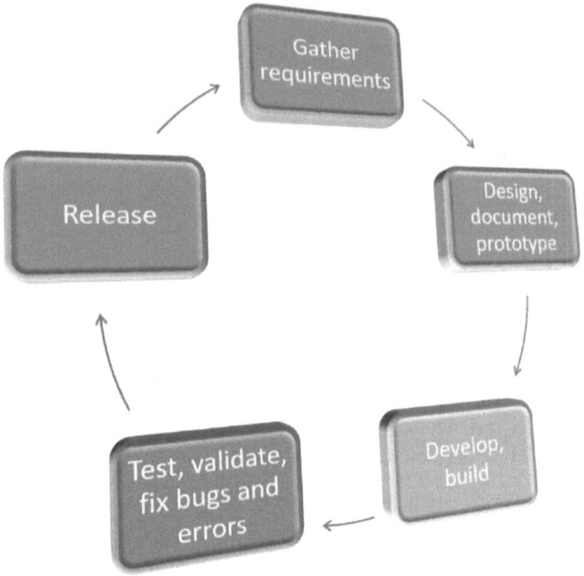

Figure 4.2 The Plan-Do-Check-Act Cycle, Used in Iterative Development

each iteration, a deliverable is produced for the customer to review, experience, and use. Feedback from the customer is used to make the necessary changes to the next component of work or the scope of the next deliverable. Through this process, continuous improvement is built into the work throughout the entire project.

Like waterfall, the iterative and agile methodologies were formally developed primarily to support software project development. In practice, these approaches are now used for a wide variety of projects. To make the world of project management even more complex, new, and similar, frameworks are cropping up all the time. Check out the Internet to discover the many assorted flavors of agile and iterative development.

4.5 Two Sides of Project Value

4.5.1 The Value of Methodology

Plenty of practitioners are self-declared purists of waterfall, agile, and everything in between. If you Google these methodologies and what works best for any project, you will find there are limitless opinions. If you are like me, you may find it fascinating that project management professionals typically focus the debate on what type of methodology works best, rather than identifying what value the methodology provides. When was the last time you heard a CEO declare that using a particular methodology is the key to delivering value from projects?

The bottom line is the organization that pays for the project generally does not care what methodology was used to deliver value on the project. Executives care that the project value was delivered as promised. Considering this, I am continually perplexed by our obsession as a profession to argue for or against one methodology or another.

Part of the problem may be that when it comes to project value, the definition is not clear or universal. We can agree that value is delivered when the project is completed on time, on budget, and within scope. These are the three components represented by the value of the Iron Triangle. It is the standard value projects are expected to produce. Managing the factors of the Iron Triangle is the first and main value that is often the focus of any project. It is clear and easy to define, and it can be measured objectively and consistently. You can control, monitor, and demonstrate this value through standard tools in the project management tool box that compare the original project estimates for budget and schedule to what occurred throughout the project. Finally, this information is collected at the end of the project to determine the overall value that was delivered through project performance on budget, scope, and schedule.

4.5.2 The Value of Stakeholder Satisfaction and Emotional Intelligence

There is another, equally important side of project value that must be delivered, which goes beyond the familiar measures represented by the Iron Triangle. Whether it is identified or not in your organization, make no mistake, this type of value is expected. It is identified, measured, and produced through the experience of the project team and stakeholders. The measure of this value is relative to cultural and environmental factors. For these reasons it is complex and not easily created using a glorious Gantt chart or a simple Kanban board. Beware: The skill to produce this type of value does not rely on logic-based intellect alone. In fact, it relies on the project manager's ability to tune in to and effectively develop emotional quotient (EQ). Also known as emotional intelligence (EI), this ability is defined as the ability to recognize, understand, and manage our own emotions, and to identify, understand, and influence the emotions of others (Institute for Health and Human Potential [IHHP], 2017). Although project managers might not talk about it much, EI is a primary tool in any project manager's tool kit. Using EI will help you choose the tools and the methodology that will produce the best project experience and value to stakeholders. The inability to develop and use EI adequately will get a project and the project manager into trouble every time.

4.6 Creating the Perfect Match

This is where timing, organization, and team environments enter into the picture. Using the right methodology can produce the wrong results if the environment is not carefully considered. The value the methodology brings needs to be matched to the team and organization's environment. Think about John and his introduction of agile tools such as the Kanban board. What do you think might happen if John introduced the full agile methodology at once and demanded the team and stakeholders abandon practices familiar to them and adopt the roles, tools, and methods of agile immediately?

The answer to this question is based on psychological studies of adaptive people. In "Why 1 in 3 People Adapt to Change More Successfully" (2016), Nick Tasler defines adaptive people as people who adapt and even thrive in times of profound change and chaos.

4.7 Purpose Empowers

Adapters are people who are able to find purpose by asking what they could do to move forward in a new environment. This purpose empowers them and gives them a method to embrace change. In addition to empowerment, there is a need to consider the individual capacity for change. This capacity for change is the combined mental and physical energy each person has which enables them to adapt and learn to embrace new behaviors, practices, and information. These energies create a personal resiliency which gets stronger in times when we feel empowered and in control. On the contrary, personal resiliency gets weaker when we feel powerless and out of control.

4.8 Timing + Empowerment = Trust

Let us go back to the question about John and review the scenario in which John requires the team to adopt agile all at once. The team is now expected to work with new terminology, new tools, new processes, new roles, and new methods, all while trying to keep the project on schedule, on budget, and within scope. On John's team, individual energies needed to adapt to change are already depleted as they try to navigate a project that is off-track. The team is now feeling out of control. By taking away familiar processes and tools without first giving the team time to adapt and master new tools, John will create more fear and anxiety, reduce feelings of empowerment and confidence, resulting in slowing the progress on the project to a crawl. In short, introducing so much change in flight on the project is a disaster. Even so, in this scenario, project failure is not the largest casualty. Consider that the most devastating loss in this story will be the loss of confidence the team and the organization will have in John's ability to lead a project team effectively and to support them to appropriately embrace and navigate change moving forward. Once the organization loses faith in John, it will be a long, difficult road back to a place where John is able to earn back the trust and confidence he lost. This is assuming the organization is willing to give John a second chance.

4.9 The Balance of Art and Science

Frameworks and tools in the practice of project management are suited for a wide variety of organizational and team cultures, project types, industries, rigor, and sizes of projects. To know how to use any of the tools and frameworks available is not just the art, nor is it just the science of project management, because you cannot find project success in a single tool or methodology. To know when to use these tools and to understand the value they will bring to your project is a delicate balance of art and science. In fact, developing the skill to find this equilibrium can literally make a difference between a project that flourishes and one that withers and dies.

For some of you purists out there, I have just stated fighting words. At this moment, you may be preparing your defense to argue how your favorite methodology is the best for every project. The internet is full of articles, debates, and blogs on this topic. Before we add our debate to this list, let me ask you once again: Does the customer, whether internal or external to our organization, care how we deliver success? I don't think so. The customer's number-one concern is the results and the customer experience along the way. Now is an excellent time for you to ask whether your focus is truly on producing stakeholder value through creating a positive stakeholder experience while on the project. Are you focused on your customer's number-one concern? If not, you should be.

4.10 Adapting for Success

Are you still wondering how you adapt tools, processes, and project frameworks to provide your customer, your team, and the organization the best experience possible while delivering results? Let's be honest, if you have been employed as a project or program manager for a while, you have probably found yourself in a position that required the use of a more hybrid approach in the application of tools and methodologies to accommodate project requirements, the organization, team culture, and stakeholder needs. Thinking about those experiences, reflect on why you altered your approach to a particular methodology to develop a hybrid. Did stakeholders require experiences or deliverables the methodology did not support? Did the team lack the willingness or aptitude to embrace the framework you were using? How much time did you spend training, teaching, and pushing the full methodology, tools, and practice that you or your Project Management Office (PMO) prescribed for the project without considering the factors outlined here? What was the result? Did stakeholders love the experience? Was the project viewed as a success? One last question: If someone gave you a magic wand and called it the methodology Timbuktu, promising that this methodology will deliver projects within the triple constraint while creating satisfied stakeholders, delighting customers with the results and the experience along the way, would you care if that methodology resembled waterfall or agile?

If you are driven to produce success for the team, your organization, and yourself, this is your challenge: Get over what the methodology or tool is called and get into what it delivers. If you are ready to take the challenge, you are ready to learn how to build your own Timbuktu.

4.11 The Timbuktu of Project Management

Timbuktu is a practice that prescribes a wide array of tools, process, behaviors, and practices designed to develop specific value for the project and stakeholder experience. I call it Timbuktu because it is a place you create for the project team and all stakeholders that provides the opportunity for all team members and stakeholders to list their needs from the experience and the project. In this place, it is your responsibility to match those needs to the right tools, processes, and way of working. If there is not an existing tool or process that fits the need, it's your job to create it. That's project management heresy, you say? While the idea might sound like a revolt against project management best practice, in fact, it is the very essence of project management. After all, this is the same type of revolution that gave birth to agile, iterative development, the scaled agile framework, and countless other methodologies and tools that help project managers deliver success today. My challenge to you is not to throw out the good work that has been done to develop these tools and methodologies. My challenge to you is to keep the innovation going! Develop what your team needs, innovate, and create.

4.12 The Magic of Timbuktu

Early in the book, we learned that project value is based equally on the experience and the deliverables from the project. To balance these factors on the project, you need to choose tools that best fit the project and the needs of the stakeholders. How do you do this? John's work with the team in Chapter 3 is a good example. Even though he was working a project to complete a complex software upgrade, he introduced the team to some of the principles and practices of the agile methodology. Through simple, easy-to-use tools, such as a paper Kanban board, John demonstrated the processes associated with agile and the tool. Over time, as the team became confident with each tool they embraced, John introduced a new level of complexity or tool, making sure he understood and could articulate the value of that tool or process to the team. As the team experienced the value of using the tool, they more readily embraced the tool and the process. Contrary to this, when a new tool or process is introduced, and the team does not understand or experience the value, what happens?

4.13 Respect Foundational Principles

Maybe now you are thinking, "I'm making this more complicated than it really is." In response, I ask you, "What is the universal reason projects fail?" If you know the answer to this question, I urge you to write the next book and share that definitive answer with the rest of us. The fact is, projects are complex and so are the reasons they fail. This is because the dynamics of work and human interactions are never simple. No matter how many times we travel the project journey, there are new people, challenges, and variables that dynamically change the experience and the outcome.

To succeed, you will need to be keenly aware of what is working and what is not, so you will be able to proactively adjust the process and introduce new tools to address the present and future needs of the project. Your ability to do this is effectively your value proposition for the project, stakeholders, and the team. Before I explain how this can be done, let me be clear on one main point. I am not suggesting in any way that you adopt an iterative framework like agile and work it in a waterfall-like linear fashion. Whether you are creating a hybrid or using the pure methodology without alteration, the guiding principles of each methodology must be respected as you apply that respective methodology to preserve the distinctive value of each practice. For example, one of the key values of waterfall is the order and predictability of the work and the heavy documentation produced along the way. One of the primary values of agile is the iterative approach that is used while creating minimal documentation. Remember, if the processes that produce the value are not respected, the methodology itself has no purpose. If this seems obvious to you, why do we continually see agile projects worked in a waterfall manner? You know what I'm talking about. I'm talking about those projects that are called agile yet require countless stakeholder meetings, project governance and rules, endless sprints that produce nothing, and piles of project documents.

The magic of Timbuktu avoids these pitfalls. It gives you the ability to create that place where you can earn the maximum value through the use of tools and practices while respecting the guiding principles along the way.

4.14 Agile in an Imperfect World

The company OZ could no longer wait to replace the antiquated set of spreadsheets and databases it had been using to manage the organization's investor program for the last 20 years. To solve this problem, the organization decided to develop a robust application that would automate all processes and transactions for the investor program. Implementation meant transferring historical data that did not meet the current standards or formats due to changes in regulatory and program requirements that had been made throughout the history of the program. OZ did not have a central data store, and they were not in the business of developing applications or technology for consumers. OZ was a supply management and marketing organization.

To tackle the need for organizing and managing several projects that would be required to prepare the data and the technology for the new application, a program framework was developed. A program structure helps to develop and manage related projects, subsidiary programs, and activities in a coordinated manner to provide benefits and control not available through managing the projects individually (Project Management Institute [PMI], 2017a).

Soon after OZ approved the program, Claire was assigned as the program manager. After reviewing the initial program charter, she recognized that using an agile methodology would reduce risk on some of the key projects because of the benefits the iterative learning cycle would provide to the team. Incremental product development would also present the best opportunities to manage the build of the new application. The build was the largest of the projects in the program. Still, there were many questions on how the overall program should be managed because of several knowledge, technical, and business architecture gaps that existed in the organization. In addition, business rules and infrastructure at OZ were changing rapidly. Stakeholders were not familiar with agile in the organization, which presented the risk that using agile would slow down or stall project work. At the same time, Claire knew that while a fluid, flexible process would be needed by the program team to keep up with the changing direction of the organization, Claire would need to be mindful of imposing too much change on stakeholders, who would be experiencing the biggest operational change in 20 years as a result of the changes the program would implement.

4.15 Program Initiation: The Start of the Journey

Claire worked with the PMO and supervisors throughout the organization to secure the core program team, many of whom would also be working some of the projects in the program. This core team would remain engaged throughout each phase of the program, from the initiation of the program and its projects until the close at the completion of the last project. Other subject-matter experts who would provide intermittent support through their expertise were added to the extended team. A steering committee was formed from the organization's executive level, consisting of the CIO, CFO, and other directors, to represent the departments that would be directly involved in the program. The strategy, scope, and deliverables at a very high level had already been determined by the program's sponsor, the CFO.

Monday morning, Claire facilitated the program kickoff. Core team members, the steering committee, and extended team members were present to review the information in the program charter and the RACI chart. A RACI chart (the acronym stands for Responsible, Accountable, Consulted, and Informed) outlines the project roles and levels of responsibility, with accountability for each role and project activity. It also indicates whether a particular role should be expected to be informed or consulted (see Figure 4.3)

Claire reviewed her role and responsibilities with the group to be sure the steering committee and the sponsor had the same understanding and expectations of Claire. Finally, she reviewed the resource requirements and commitments from participating departments, asking for a commitment from supervisors and directors to protect each team member's time, so they could complete project deliverables. The information she shared was aligned with the information on project roles and responsibilities in the *PMBOK® Guide*, Sixth Edition (Project Management Institute [PMI], 2017a) and the PMO at OZ.

Some of the participants in the meeting were not familiar with the concept of program strategy, decision making, project sequencing, and management, so Claire reviewed those components with the group and explained why they are important. Confirming the program strategy and getting agreement on the process that the steering committee and program team would use for decision making was essential before the program began. Subsequently, the group agreed to a process that would help manage the group when they could not agree on the

Program RACI: Investor Oz — RACI Diagram

Function / Activity / Deliverable	Sponsor(s)	Project Manager	Product Owner	Business Analyst	Core Team	Communication Lead	Extended Team	
Initiation								
Complete Needs Analysis	A	R	C	C	C			
Prepare Business Case	A	R	R	R	C	I		
Prepare Cost/Benefit Analysis	A	R	C	C	C			
Conduct Risk Analysis	A	R	C	C	C	C		
Prepare Charter	A	R	C	C	C	C		
Seek Sponsor and Board Approval	A	R	C		I			
Obtain Funding	A	R	C		I			
Define & Plan								
Create baseline schedule	A	R	C		C	I	I	
Assemble Core Team	A	R	C		C	C		
Prepare Communication Plan	A	R	C		C	R	I	
Prepare Project Budget	A	R	C					
Define Legal & Regulatory Requirements	A	R	C		C			
Gather Requirements	A	R	C	R	C			
Establish System Designs	A	I	C	R	C			
Prepare Training Plan Going Forward	A	R	C	R	R	C	I	
Prepare Post Project Support Plan	A	R	C	R	R	C	I	
Execute & Perform								
Purchase	C	I	R		I	I		
Install	A	R	R	R	R		I	
Monitor & Control								
Control Schedule	A	R				I		
Report Performance	A	R	C		C	I		
Monitor and Control Risks	A	R				I		
Manage Project	A	R	C		C			
Budgeting	I	R	C		C	I		
Scheduling	A	R	C		C			
Procurement Planning	A	R						
Quality Management	A	R	C		C			
Close								
Complete Lessons Learned	A	R	C	C	C	C		
Send Satisfaction Survey	A	R				C		
Project Closeout	A	R						
Celebrate Success!!	R	A	R	R	R	R	R	R

R = Responsible for performing the work; the 'doer', can be shared
A = Accountable, the person who is held accountable that the action/task is completed; should be one per activity
C = Consult, someone who has input
I = Inform, someone to be informed of decisions or results

Figure 4.3 A RACI Chart

direction, prioritization, or objectives for the program and its relative projects. Even though Claire made every effort to clarify this information in the program charter, she knew that as the team moved forward and the group became more knowledgeable, they would be challenged by changes resulting from the nature of the effort and the rapid environmental changes currently happening at OZ.

At the end of the meeting, the intended objective was reached. The group had a clear high-level idea of what was needed to develop the program and its deliverables. Together they developed a unified vision on how they would work together, make decisions together, and keep their commitment to the program.

4.16 Program Planning

Claire tackled roles and responsibilities in a separate meeting with the core team, to make sure they understood the roles of scrum master, product owner, project manager, and business analyst, as well as the roles the subject-matter experts would fulfill in their respective disciplines. Claire was now responsible for engaging the core team to identify the projects and drill down the work required to complete those deliverables. She knew she needed to be flexible and thoughtful with the team in defining the tools, framework, and process for each project, to assure that the mix would provide the best value and experience. This method is consistent with recommendations in the *PMBOK® Guide*, Sixth Edition (Project Management Institute, 2017a), so Claire felt confident in her approach.

The program timeline was approved for 36 months. Claire knew the information outlined in the program charter in terms of scope, schedule, resources, and even projects was likely to change, because the core team responsible for identifying and estimating the work were not involved in those estimates. The first order of business would be to get the core team together and begin to break down the work. She needed to find a straightforward way to build collaboration that would inspire this new group of experts to understand the work and each others' perspective of the projects in the program at a high level.

To start the journey, Claire chose to build a work breakdown structure (WBS) with the team. A WBS is a visual high-level depiction of the work in large buckets (see Figure 4.4). Using this tool, the team would work to identify the projects needed in the program, then the major deliverables for those projects, and what would be needed to complete them. Claire chose to break down the WBS by deliverables that needed to be produced through the program. To get the brainstorming going, the team was asked to write underneath each deliverable that was listed in the program, the components or buckets of work that would need to be completed to produce that deliverable. Soon after they began the exercise, the team had a clear idea of the scope involved to develop each deliverable. By using the WBS as a tool to identify the big buckets of work and then define the scope surrounding that work, the team was able to envision and develop an outline of the projects that were needed in the program. Organizing large deliverables by project enabled the team to focus each project on exactly what was needed for the work. No project in the program was longer

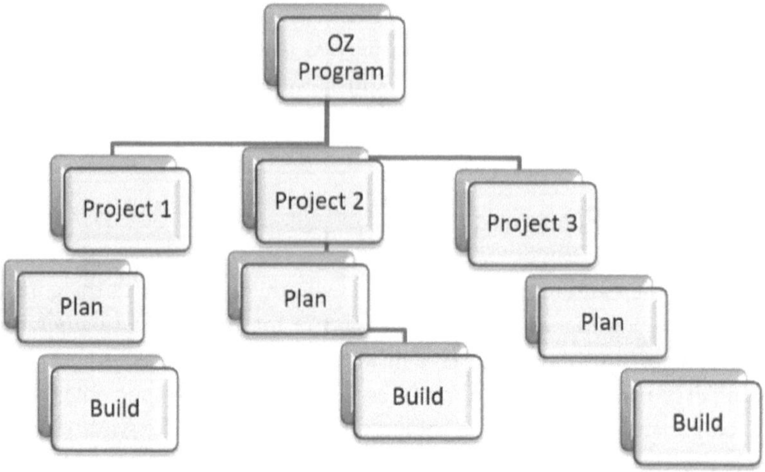

Figure 4.4 A Generic WBS

than 18 months or shorter than 3 months. The team also created a WBS dictionary to clarify jargon or industry lexicons so everyone on the program was speaking the same language. A WBS dictionary is a document that provides detailed deliverable, activity, and scheduling information for each component in the project (see Figure 4.5).

By the end of the first week, Claire and the team were well on their way to discovering what needed to be done at the program and project levels in more detail. To get even more detail, Claire decided to engage the product owners and the core teams in a word cloud activity. Each project deliverable and related work package from the WBS was placed in its own "cloud" and displayed on a poster in the team room. The group was gathered together and asked to individually write on a Post-it® note 6 to 10 words that describe the buckets of work they believe would be needed to produce each deliverable. Then Claire invited the group to place their stickies on the poster board. Claire knew that this activity would help the team interact with each other while getting them engaged to define the work. While the team was interacting, Claire entered their feedback into a word cloud template. By the end of the session, the team had flushed out more detail for each deliverable and understood the larger pieces of work. Claire displayed the online word cloud (see Figure 4.6).

The word cloud template allowed the team to see the words that were most frequently mentioned during the exercise. Although the exercise was primarily an engagement activity that could not replace the process of gathering requirements, it showed alignment in what the team was thinking and highlighted items that might represent the largest chunks of work.

Work ID	Work Name	Description of work	Completion Criteria- Quality, Timelines, Quality	Customer External/Internal	Asset Type/Resource	Priority level for work (level 1 -3)	Estimated Time (start to finish)	Skill Required (E, S, N)	Work package/deliverable Dependency 1.0 Before 1.1?	Who is Responsible for work	Party responsible for work package or deliverable approval
1.0	Plan	Plan the project, identify requirements, document information.	Quality: To the level the core team can adequately plan and execute work	Internal business department	Core Team	1	6 months	Expert	Before 1.1.	Project Manager	Internal Business department
1.1											
1.1.1											

Figure 4.5 Format for a WBS Dictionary

Figure 4.6 The Word Cloud Created by Claire and Her Team

4.17 Creating Strong Performance

At this point in the program planning process, Claire had presented an array of tools to the team. She used these tools to engage the team, create collaboration, and collect information. Claire was also able to observe the team's acceptance of each tool and process, asking for feedback after each experience. Claire engaged with the team in this process to assess how well the team adapted and worked together. By observing them working together in the sessions, she was also able to identify the dominant, extroverted, and introverted personalities on the team. The process allowed Claire to identify personality conflicts and observe how each individual team member and the team as a whole managed conflict.

Claire knew the value of these exercises was to help the team to move through the forming process, transitioning them from a group of individuals with different goals and ideas to a united group with a clear, unified goal and vision. To accomplish this, Claire needed to understand each team member's personality and their capability gaps, so she could identify what training or support would be needed to move them quickly along the stages of team performance, starting with forming through storming, into norming, and finally to performing. Performing is the stage of team formation where the team achieves optimum performance. Claire's strategy was to continue to use tools and processes that created opportunities for engagement, empowerment, and collaboration within the team while enabling them to further define what would be required to complete the program.

Over the next few weeks, Claire developed familiar patterns within the team in the way they worked. Four consecutive days a week the team met, and while standing, gave their updates on what they completed yesterday, what they will do today, noting any concerns or barriers. Claire listened while each team member shared their update, picking up on verbal and nonverbal cues that might translate into action items for Claire to remove barriers or address concerns which might slow or derail individual progress. Through this process, Claire helped each member remain unhindered and able to focus on their work.

Engagement, leadership, and individual growth of each team member was also a large part of Claire's work with the team to help them strengthen their individual skills. To do this, as the team became comfortable, Claire asked each team member to step up and lead the stand-up meetings. This is a core responsibility of a scrum master, a role typically filled by the project manager. At the same time, she did not hand off her scrum master responsibilities to remove impediments for the team. Claire was very strategic in selecting this division of responsibility for the scrum master role, allowing her to provide leadership opportunities for each team member while preserving the team's ability to stay focused on their work.

4.18 Empower and Engage to Innovate

This was the largest and most rigorous program the company had approved, so the team knew they would need to develop new tools to help them succeed. Claire's approach was to engage every role on the team, encouraging them to present the case to use existing tools, create new ones, or change the process around a familiar tool, making sure the expected value and risks were carefully examined by the team for each case.

Claire met with the product owner, Mary, and the business analyst, Tom, to discuss how they might define a process to engage extended stakeholders who would be using the new application. Some of the stakeholders had been hired to set up the user access interfaces 20 years ago. For these stakeholders, their fear of this change was fueled by countless unknowns. The goal was to find a way to engage these individuals efficiently in the project and the development of the application to create transparency and diminish the dreaded unknown. Claire had tools for this in her project management tool box. She knew that an agile framework provides a means to state the testing criteria for each story the team would develop once implementation began. Claire explained to Mary and Tom the need to develop a consistent method to engage these stakeholders, reducing unknowns, fear, and resistance while increasing acceptance of the many changes that would be required throughout the project. Claire introduced them to a test-plan template and an end-user feedback form, explaining the benefits and gaps of each tool, which demonstrated that neither tool completely met the challenge. She then asked Mary and Tom to work together for the next week to create a tool and process that would satisfy the project's needs. The following week, the group met again. In this meeting, Mary and Tom presented the new end-user feedback form and process. The tool and process they presented met the project's needs while minimizing risks. Mary and Tom were eager to champion the process and use the tool. Tom offered the objective analysis to measure the effectiveness of the tool and process, and Mary provided guidance needed to assess the impact of the changes on the staff who would use the new application. Soon the two were fully engaged with stakeholders well before the team began developing the new application, consistently measuring feedback and the effectiveness of the new feedback tool and process. Quickly, they learned what improvements might benefit the feedback process based on end-user input. This allowed them to make refinements throughout the process until it flowed like clockwork. Claire followed this method with all the core team members, challenging them, engaging them, and empowering them to create and innovate as required to develop the tools and processes needed to support project success, create transparency, promote collaboration, and reduce fear. If a standard tool from the project management tool box fit the need, Claire introduced the tool and the value at the appropriate level of rigor. Selecting the appropriate tool and applying the right level of rigor was a crucial step which spared the team from applying more effort and time than necessary while ensuring that the team's efforts resulted in immediate value for their organization.

When the core team reached the end of the planning phase for the first project, they fully embraced their role as a leader, each in their respective discipline on the team. Team members sought opportunities to innovate, creating new transferable tools that provided sustained measured value for the project and the organization. As a result, each professional on the team flourished, learning new leadership and technical skills along the way.

4.19 The Value of Timbuktu

As time went on, Claire and the team suffered through some of the same challenges and gaps typically encountered by a program of its size and complexity. The difference in this team was demonstrated through their behavior when those typical challenges arose. The team, clear on their purpose, empowered, and confident in their capability to create workarounds to mitigate or remove the challenge, thrived. They kept a steady pace throughout the duration of the program and did not lose focus when they encountered surprises or challenges along the way.

At the end of the journey, the organization and the team understood that Claire's ability to build an innovative, efficient team environment was her true value to the organization. Claire successfully created Timbuktu for the team.

4.20 Your Challenge

What type of environment do you create for your project or program team? Do you have the ability and the courage to create Timbuktu, a place of innovation and empowerment?

As you reflect on these questions, remember, it is not our knowledge that creates our experience. It is our experience that creates our knowledge. Lead the way. Create the experience. Be the value your organization and your team need to succeed.

If you are one of the more fortunate project managers who live in a world where the same exact work is completed to produce the same exact product for the consumer, I challenge you to define that what you are doing is really a project. Remember, a project is a temporary endeavor to produce a unique product, service, or result. This definition may seem clear and obvious to you. Frankly, it's obvious to me. Yet, in the business world, it's mind boggling how often this definition is misunderstood. As a result, there are project managers using the tools and skills of project management to manage operational work with the sole directive to "herd the cats" in order to organize tasks and the people who are doing them, even though the deliverables and the work they are doing to create these deliverables are not unique. I have met a few of these project managers, and I bet you have too.

The good news in these scenarios is that the business recognizes the value of project management, the tools, methodologies, and processes it brings to the table. The bad news is that the project managers are often set up to fail because they are used as cogs in the wheel to speed up the delivery of the work and monitor the performance of staff.

To think about this in another way, decide which scenario best represents project management: a row master cracking the whip in the hold of an ancient Greek warship to keep the men on the oars rowing faster, or a master builder in the shipyards organizing and directing the work to build the Greek warship.

How can we expect others outside of project management to understand the value of project management and how it best serves the organization unless we as project leaders demonstrate the value every day? Each of us must be diligent in our efforts to guide others on the best utilization of project management to protect the value of the profession.

Frankly, what is and what is not a project and productive use of the practice can be a bit ambiguous, because the tools, processes, and behaviors that are used to deliver a new service, product, or result to the organization are equally good for managing day-to-day work. It is all about leadership. The hard part is not using your skills, the hard part is using what you know to be a leader, every day.

4.21 Just Crazy Common Sense

About now you are either ready to throw this book out the window or you are scratching your head trying to figure out how the innovation of tools and methodologies can be a good thing for you or your project. After all, there are plenty of articles that share stories of disastrous failures because project managers created hybrids, morphed tools, or blended practice between methodologies and frameworks on a single project. So how can innovation be a good thing?

The fact is we live in the real world, where organizational change cannot be forced. Even the best PMOs have to adapt to the requirements in their environment, which may mean bending the tools or methodologies a bit. In the real world, we do not control many of the environmental and cultural factors that affect projects and teams. To provide the value our organizations are expecting and the value that has been promised by the profession of project management, we need to be flexible, adapting our approach and the tools and processes we use to fit the needs of the organization and the environment. How do we accomplish this if we follow a templated approach to waterfall or a templated approach to agile?

This idea is now articulated throughout the *Agile Practice Guide* (Project Management Institute, 2017b), published by PMI and the Agile Alliance. This guide is included as a supplement to the *PMBOK® Guide*, Sixth Edition (Project Management Institute, 2017a). Times are changing in our profession. Are you changing with them?

The threat to successful innovation of our approach, tools, or methodology is not the innovation itself. The threat to success is applying innovation to create a hybrid in a tool, practice, or methodology without thoughtfully evaluating the benefit and the risks first. Claire's story shows this journey. She knew agile would be the best framework for the project. She also knew that careful identification of the many technical and business infrastructure gaps would need to be completed so her team could work to evaluate the impact of those gaps on the project first and then work to close gaps that might prevent successful implementation of the new application. To complete this journey, Claire organized the initial planning using some of the traditional waterfall roles, tools, and processes. Along the way, she exposed the team to agile methods they would be using in full force during the project. Finally, she mentored and coached her team to be the catalyst for continuous improvement of the tools and practices they used throughout the project.

Agile and waterfall are not perfect. Neither are the environments where we try to apply them. Do not strive for perfection. Strive for value. For every practice, tool, and action you use, take the extra step to define clearly the value and the risk. Be the coach and mentor to help others follow this path and you will prove your worth, creating the place called Timbuktu for your team, and delivering the value promised by our profession.

Chapter 5

The Magician and the Disappearing Magic Wand

© Jim Kangas

As we discussed in Chapter 4, the profession of project management promotes the benefits of signing on a project manager to increase the odds that a project will come to fruition as originally planned, delivering value with all three sides of the Iron Triangle intact. The proposition that value can be delivered with all original sides of the Iron Triangle intact can seem daunting, almost impossible, in the face of the unpredictable change experienced by industries and markets every day. Combine unpredictable changes to business requirements with stakeholders who have personalities similar to Sybil, Margaret Mead, Einstein, and Donald Trump, you will need more than a grip on scope, schedule, and cost to deliver success: You will need to become a magician!

If you create Timbuktu, you might have a chance to produce some of this magic. Timbuktu is the environment we described briefly in Chapter 4 that you create for your team, which promotes the team's ability to adapt, create, and innovate tools, methods, and behaviors that will promote success for the team and the project. There are two main components that you need to create Timbuktu: a project room or space the team can claim as their own, and a productive team culture.

5.1 The Space

This is a space that has room for one-on-one as well as group discussions. Technology supports this space so that everyone in the team can jot ideas down on their own whiteboards. The team also has a central space for jotting down ideas, sending screenshots over email, and creating visual demonstrations that the team and other stakeholders can view on a large screen. The space is convenient and centrally located. It is well lit with natural and artificial lighting. Team members have workstations in the room that are ergonomically designed and offer support during long, intense hours of work. Basically, it is a place that is comfortable, well equipped, and pleasant to occupy, which promotes efficient completion of work.

5.1.1 The Virtual Question

Before we dive into the second item you need for Timbuktu, I want to address the question of virtual teams. At this point, you may be thinking, "I don't need to create this place, we can create it virtually." Let me caution you on this assumption.

In today's world, you can create almost anything virtually except trust, commitment, excitement, dedication, and—let's face it—interpersonal relationships.

While you can meet and talk to anyone virtually, you cannot create a lasting, productive connection. The words that are spoken in a conversation may resonate with you, yet without all the other components that connect us to another human in a way that promotes trust, honesty, and collaboration, you are not able to fully connect the communication with the speaker's intent. Those components include the ability to observe a person's behavior, body language, and response to the environment and people around them. Eliminating the ability to observe communication cues such as body language and facial expressions reduces the effectiveness of team communications. Numerous studies confirm that only 7 percent of our communication occurs through the words we use. The rest of the message is comprised of nonverbal communication, body language, and tone (Knapp et al., 2014). Have you ever been told you need to have a "poker face"? I have been told I need to get better at my "poker face" because, apparently, my emotions are written all over my face! Even when I do not utter a word, people in the room are assessing my opinions of the topic being discussed based on my facial expressions and body position. Imagine the impact on the message when you receive only 7 percent of the message! When that occurs, we need to figure out how to interpret the message with less than 10 percent of the data. What does that mean? Chaos would be a certainty or at least a very painful version of the telephone game requiring each participant to volley back and forth the same message in an attempt to clarify its meaning. Does this sound like something that has happened to you while trying to communicate through email?

Since the 1990s, virtual teams have become a global necessity to overcome time, space, and cost barriers for organizations. Today, the use of virtual teams is now a widespread practice in almost every business throughout the world. With the passing of each decade, the technology that has been produced helped to close communication gaps in the virtual world. Today, Skype and FaceTime help us pick up on those nonverbal signals by offering us the capability to view in real time people or groups as we carry on a conversation. Virtual communication that includes real-time visual screens improves our ability to pick up on nonverbal cues, but it cannot provide the full 93 percent of information we need beyond the words that we use to interpret the message correctly.

Focus is another problem with virtual communication. To attempt to do more with less, how many times have you been guilty of working on your computer while also participating in a virtual meeting that had nothing to do with the work you were doing on your computer? I know I am guilty. Later in the book, I'll explain more about the impact to our productivity when doing two or more things at once. For now, let's just say there is scientific evidence that the practice of doing more than one thing at a time does not benefit you, it does not benefit the group you are meeting with virtually, and most important, it does not set you up for success.

5.2 Culture: A Prerequisite for Health

The second component you need to create Timbuktu is a healthy culture. Teams develop their own culture over time. Team culture consists of social norms, behaviors, and customs that are woven into their daily activities. A healthy culture is created by developing behaviors that promote an environment within the team where honesty, transparency, trust, and the ability to respectfully challenge the status quo are the essential cultural norms. In your leadership role with the team, the creation of this type of team culture falls in your wheelhouse. To be successful, you will need to embrace all the behaviors outlined in Chapter 3 of this book.

5.3 The Magic Wand

Creating Timbuktu will generate some of the magic you need to develop high-performing teams, yet this alone still won't be enough to produce sustainable value for your organization. So, how do you find your magic wand to produce this kind of success?

In the world of magic, wands were once a popular tool for most magicians. The wands produced the illusion that the power of the magic was coming from the magician's tool—the magic wand. Today, wands are disappearing as props in the magic show and the focus is more on the magician. This creates more of a challenge for the magician, because the wand can be used as a distraction for the eye, taking the focus off the magician and what he is really doing to produce the illusion.

Even without the wand as a prop, magicians utilize a variety of tools and processes to amaze and delight audiences everywhere. Tools used by modern magicians such as David Blane and others rely on technology and advanced engineering. To complete each trick successfully, the magician and a team of experts construct sets, develop and use new tools, and create props, in preparation for the show. Hours and sometimes weeks of practice are completed by the team to ensure that the timing of each prop, tool, and action on stage is synchronized. At show time, the magician and the tools come together in perfect harmony to produce the desired results of amazing and delighting the audience.

The truth about magic is that it does not defy reality, it uses tools and processes to show us reality differently than we have seen it before. That is why we call it magic.

In this chapter, think of the magician as a project manager. The wand represents the tool that is used as a constant side kick, relied on and proclaimed to be the source of the magic. In this analogy, the project manager's wand might be a

Gantt chart or project task list. As we explore this concept for project management, try to determine whether it is the wand or the magician that generates the magic for the teams you lead.

5.4 Z-Nobel and Constrained Resources

Magic Trick #1: Reappearing Resources

Three months ago, the project plan for Z-Nobel was reviewed and approved by executive management. Today, due to the need to complete other priorities in the organization, Sue and Joan, two of the team's technical members, have been asked to spend less than half the time that was originally outlined for them to spend on the project in the approved project resource plan. The work assigned to Sue and Joan is part of the critical path on the project and cannot be cut from the project scope. This means all the work that needs to be done by resources, Sue and Joan, must be completed or the project cannot succeed. When Dan, the project manager for Z-Nobel, heard the news that Sue and Joan's time on the project was reduced, he quickly worked through a few alternate schedule scenarios based on the new resource limitations. After a couple of hours, Dan created views in a Gantt chart for each scenario, manipulating the order of tasks and the time to do them to predict which scenario might help keep the project on track despite the reduction in hours for Sue and Joan. Yet, no matter how Dan tried to allocate the work to other team members or reorder the work in the schedule, all the scenarios showed that with the new reduction in resource allocation for Sue and Joan, project completion would be pushed out six months to a year beyond the original two-year deadline for the project. Dan thought, "Cutting Sue and Joan's allocation to the project adds additional time of 50 to 100 percent to the schedule. This means the resource reduction will also impact the project budget because we will be working six months longer than originally planned. I bet no one has given thought to how this will reduce the payback for the product we are delivering either. That means the impact would create financial implications and tie up the rest of the core team too. The resource constraint that is being caused by the reduction in Sue and Joan's time will also impact the other projects that are scheduled to begin when Z-Nobel completes, because these other projects are dependent on the experts from the Z-Nobel core team. Wow, I have some solid facts here. I'm sure that given these facts, Kyle will change direction and help to keep Sue and Joan dedicated to this project. It's just a matter of showing him the data."

Confident that Dan had a convincing argument to maintain the original resource allocation for Sue and Joan on the project, Dan immediately scheduled a meeting with Kyle, the CIO, for the following day. The next day, Dan arrived at the office with the information carefully prepared for the meeting. In 30 minutes, Dan was able to present the information that proved the negative impacts to the project and the organization if Sue and Joan's hours per week on the project were reduced. During the meeting, Dan watched Kyle carefully to read his body language to gauge any signals that the information was surprising or sparked concern. Kyle's reaction was calm and at times appeared almost uninterested during the short 30-minute session. At the close of the meeting, Dan was baffled to discover that despite the solid data presented to Kyle, he was unable to convince him to provide funding for contract resources, allocate more internal technical resources, or keep Sue and Joan allocated to the project as originally planned. Kyle ended the meeting with Dan by giving him this challenge: "Dan, I'm confident you can figure out a way to stay on schedule with the project. After all, we pay you project managers to manage these types of challenges, don't we?"

Dan could feel his face getting flushed. He clenched his fist while responding, "Sure Kyle, I get it. You need me to pull a rabbit out of a hat."

"That's it, Dan! You've got it. I have confidence that if anyone can do this, you can Dan," Kyle said smiling as he left the room.

Dan went back to his cubicle frustrated and confused. "How can he put all this on me?" he thought. "I can only deal with the realities I'm given to work with. This isn't fantasyland. How can Kyle not see the damage he is causing by cutting the time Joan and Sue can spend on this project? The information was so clear. I put all the facts right in front of him. It's almost like he does not see the same realities I see. What could be more important than keeping this project on schedule, given the costs that would hit the organization if it slips?" At that moment, Dan suddenly became aware that his reality and priorities were different from Kyle's. If he was going to succeed, Dan knew he needed to understand Kyle's reality and his priorities. Dan thought, "Hmm, maybe a different approach and a bit of resource magic will do the trick?"

The next step was to meet with Sue and Joan's manager, Josh, to discover more options they could uncover together to keep the project on track. During the meeting, Dan realized that two other internal developers were working on a different internal technical effort that really fit the job responsibilities of a database administrator (DBA). The responsibilities on the other effort were consistent with the expertise normally required from a database administrator, including designing, implementing, and maintaining the database system for another priority information technology (IT) project. In addition, the scope of work for the effort did not require development. Once Josh and Dan really dug into the requirements for both the internal effort and Dan's project, it became clear to Josh and Dan that these IT development experts were underutilized as they worked to fulfill responsibilities for a DBA.

Dan knew the next step was to find out who could take over the DBA responsibilities to free up the development resources, so the developers could then assist with the scope of work on the Z-Noble project and fill the gap left by the reduction in hours for Sue and Joan. The goal was to balance resources on both efforts while providing the expertise that Dan needed on the project. As manager of the two developers acting as DBAs, Josh was onboard with this approach too, because it might provide an opportunity to cross-train his staff while stretching their capabilities and building their expertise. For the next week, Dan and Josh worked with the resources in IT to define a plan that would meet all internal IT needs without slipping the Z-Nobel project. By the end of the week, Dan felt prepared for his next meeting with Kyle. This time, he prepared the view of reality that Kyle wanted to see. Dan realized that Kyle's primary concern was not how the projects get completed on time. In fact, Kyle had little interest in how the projects were completed.

Dan worked with Josh and the developers to build out the timeline for the IT efforts they were assigned. They were careful to examine the Scope of Work (SOW) that each resource was assigned on every effort that required their expertise. The idea was to sequence the work so each resource was sharing their skills, transferring knowledge, while serving multiple roles, offloading nontechnical duties to administrative support, and slashing nonpriority meetings from the schedule. This technique helped Josh build the skills within his technical team, because it provided opportunities for staff to focus on learning and transferring their knowledge when it made sense without negatively impacting their time. In other words, skill transference was done when the benefit of doing so was greater than the investment of time the resources spent on sharing their knowledge.

For example, if Sue estimated it would take four hours to get Beth, another internal developer, ramped up on a segment of development for Z-Nobel, the four hours would be built into the project schedule provided the return for Sue's efforts gained more than an additional eight hours of productivity for the project. This was assuming, of course, that Beth's allocation to the Z-Nobel project would not negatively impact another effort.

Once Dan put in the dependencies and resource estimates based on work effort for each task in each project to which the resources were assigned, Dan used a scheduling tool to roll all project schedules into one master schedule. From there, the group was able to see the resource requirements clearly by day, week, month, quarter, and so on, for all efforts within a given period of time. The group spent a day defining the work for the next six-month period in a manner that assured no resources were overallocated on the total efforts they were working. Ironically, this process was very linear or waterfall-like, even though Dan knew that waterfall would not be the framework used on these projects. The project team would use agile tools and methodologies to complete the work. To sequence the resources and their efforts on all projects over time, showing the complete dates for the project graphically, Dan knew a visual timeline would be the best way to demonstrate the story he wanted to tell Kyle. The group also knew the schedule estimates would need to be reviewed and refined every four weeks, because for every month beyond the first four weeks on the schedule, schedule predictions are typically less accurate for any project. This is due to a variety of market and economic factors that impact priorities in the organization and, in turn, the resources that work them. Ultimately, the objective of the work was to lay out a roadmap with the group to understand how and when activities such as sharing expertise and tasks, transferring knowledge, training, and mentoring each other would best benefit all project efforts so the group could move forward to meet expectations.

Josh and Dan planned to monitor and control each resource in the schedule to protect the team from being pulled into other activities that would drain the number of hours they had available to work on the projects. The group spent considerable time talking about the process for transferring knowledge to each other to gain the skills to support Sue and Joan. They discussed what the process would look like and how much time it would take to transfer the knowledge and skills that were needed. They examined how long it would be until other resources would be able to complete development tasks without continuous mentoring from Sue and Joan, and what outside training or additional support would be needed to make this possible.

Finally, the group worked through the sequence of events for each project to determine how the work could be segmented to produce deliverables incrementally. Using this approach may offer an opportunity to release a working piece of the project deliverable into production earlier, creating greater satisfaction for stakeholders. Dan was cautious with this approach, however. Generally, Dan liked to play it safe when making delivery promises on projects, adhering to the stakeholder management strategy to underpromise and overdeliver. After all, these scenarios were based on many variables in the organization that were bound to change. Dan did not want to encourage stakeholders to raise the bar on expectations that already seemed impossible to meet.

At the end of the work session, Sue, Joan, and the other resources were excited about the new opportunities to grow their skills. The group became confident that the new resource plan would work. Dan and Josh felt they had done a decent job of creating a work plan for the next six months that would improve the utilization of time and expertise for each individual on all projects in the queue.

Dan often wished that the organization would allow IT resources to be dedicated to one project at a time. For years, teams and project managers in the organization presented solid data on studies on brain science and real-world project results which proved objectively the benefits and efficiencies of assigning dedicated resources to a single project. What we know as multitasking, modern science has identified as "switch-tasking," a state that can negatively affect an individual's productivity up to 40 percent (Weinschenk, 2017). Even with all the information indicating the negative impacts on performance when resources were not dedicated to a project and were instead shared by several projects at the same time, stakeholders throughout the company felt better when more things were started earlier and all at once. As a result, the organization was unwilling to move to a dedicated resource model for projects regardless of the data.

Finally, Josh and Dan were ready for the big meeting with Kyle. The goal was to present the information in a way that Kyle would receive the message without preconceived assumptions. Josh and Dan knew the answer to getting the message across to Kyle was to focus on what results would be produced by using the new approach. The strategy was for Dan and Josh to avoid describing in detail how the results would be produced. After all, seeing the results from the magic and not the process that created the results is the allure of the magic trick. You see the results, not the process. Quite simply, as a member of the audience you only want to see the magic! The same is true for most project stakeholders and sponsors who are not involved in doing the work on the project. This time, when Josh and Dan met with Kyle, they laid out a timeline that showed what deliverables could be released per quarter for the next six-month period. When they presented the information, Dan noticed that Kyle was pleased with the timeline and the results. At that moment, Dan and Josh made their pitch to Kyle. Dan stated, "Kyle, we believe we can produce more without allocating additional internal resources to these projects and without contracting staff. Here's the catch: We need you to trust us to manage the resources we need directly. Protect our ability to secure the time for the resources we need on these projects, and we will deliver. What do you think?"

Kyle sat silent for a moment, then leaned back and said, "If you can deliver more, sooner, with less or the same level of staffing, what do I have to lose? Dan, you have a deal. Make it happen and don't let me down."

When Dan and Josh left Kyle's office, Dan knew he was taking a calculated risk making this deal with Kyle. If unforeseen events affected the staff that were needed to complete the projects, Dan might not be able to deliver as promised on Z-Nobel. There was no certainty of failure or success with the scenario Josh and Dan presented. There was opportunity for new and greater success that would earn benefits and satisfaction for project stakeholders and the organization that would last well beyond the project end date. The risk of failure paled in comparison to the certainty of failure in the original scenario, when Sue and Joan's time on the Z-Nobel project was reduced by 50 percent. In that scenario, failure was a certainty. By working with Josh to create an alliance, Dan strengthened his ability to manage resources effectively and protect resource time to assure the best utilization of expertise across projects and the organization.

The new direction required that Dan invest an extra week to complete the additional steps needed to gain Kyle's approval for the new resource plan. Six months later, the magic Dan performed yielded the returns as promised to the organization. Dan did indeed appear to pull a rabbit out of a hat.

5.5 Project Zephyr—Scope or Schedule?

Magic Trick #2: Disappearing Scope Becomes Schedule/ Resource Management

Rick has been working with a project team for a year on a two-year project to develop an application that will allow customers to access their investment account in the private enterprise Blue Run. The original scope of project Zephyr included building an online portal and a mobile application, so customers could access their information anytime, anywhere, from any electronic device. The team was composed of a group of experts in their respective fields: development, database administration, business analysis, project management, and others. The project used the agile scrum methodology and they all understood their roles. The organization felt confident that if any team at Blue Run could complete a project on time, on budget, and within scope, Rick's team was the one.

Six months before the launch date, the team on project Zephyr determined it was unlikely they could complete the full scope of the project without additional resources. The original charter called for a dedicated team to complete project deliverables on a specific schedule. Although the organization agreed to this, other business priorities required that the technical resources be pulled off the project from time to time throughout the last year. This affected the schedule, which delayed the release of key deliverables, risking that the project would not complete as scheduled without additional resources. After a brief meeting, it was determined that Rick needed to find another way to launch the new application on schedule. Rick knew this was his opportunity to demonstrate the value of a project manager, support the team, and help Blue Run deliver the application to its investors. He also knew this was the point in the project where the work he had been doing to engage the product owner and create a sense of ownership for the project would really pay dividends.

Rick scheduled a quick check-in with Karen, the product owner for Zephyr. As product owner, Karen was ultimately responsible for maximizing the value of the final product that would be developed as a result of the project. Essentially, the team looked to Karen for direction on the functionality and detailed scope of each deliverable, because Karen was accountable for the outcome of the deliverable (Project Management Institute, 2017). "Karen, the team is working hard on development of the application, and you have been an excellent product owner," Rick began the conversation. "I'm confident that together, we have done everything we can to assure success, now it is time to do more. We have been denied additional technical resources which we need to stay on schedule. I've examined the remaining requirements which we put together as user stories in our backlog and worked up some scenarios based on our current velocity, which shows us the average rate at which the team is able to complete tasks. The information points to the same conclusion every time. If we cannot increase our technical resource time on the project, we need to reduce the work and cut out some of the scope of the project. I've prepared a view of the deliverables we promised to complete in the original charter, so we could see each of them. I've also represented them by the quarter in the year we estimated we can release each deliverable. Based on the level of effort we have estimated for each deliverable, it looks like there are now three deliverables that will complete beyond the project end date."

Karen sighed, "Rick, I'm with the team every day and I see how dedicated they are to complete all deliverables of the project. I also know that other business priorities have pulled away the project technical resources. It's clear you've done everything you can. The problem we have is Blue Run has been telling our investors that the application would be delivered as scheduled. They are expecting the rollout as promised. If we don't roll out as promised, I'm afraid it would shake investor confidence."

Rick responded empathetically, "I understand, Karen. Looking at the information, I think we have an opportunity to deliver the application as promised. The key to realizing this opportunity is to look at the effort these components displayed in our diagram require. For example, giving outside investors the capability to access the application on any mobile device anywhere requires a significant level of effort. Did we promise that capability to investors as a feature of the new application, Karen?"

"Now that I'm thinking of it, we didn't."

Rick stated enthusiastically, "Do you think you could identify other components or functionalities that you would consider noncritical for the new application?"

"Sure, I can do that, Rick, but how will that help us deliver the application on time?"

"If we can pull out a few features from our scope to identify and prioritize those features and functions that are absolutely required to give the minimum acceptable level of quality and functionality to our investors from the new application, I think we can assure on-time delivery" said Rick. "As product owner, the choice is yours. What do you think?"

Rick knew that by presenting the option to cut scope as Karen's choice, he was respecting her role and empowering her to make the choice and own the outcome while at the same time supporting what she wanted from the deliverables produced by the project team.

"OK, Rick, I think I understand where you are going. If I work with you on this, how do I get the nonprioritized features and functions added to the application? Also, won't adding them later require more work or re-work from the team?" Karen questioned.

"Not if we do it right, Karen."

Karen shifted a bit nervously in her seat and said, "OK, I'm willing to give it a try."

Rick stated confidently, "Great, let's get started."

Karen worked through the information with Rick quickly to identify the functionalities that could be removed from the initial release of the application. They pulled the rest of the team together to examine the new, reduced scope, and asked for feedback on any dependencies or technical requirements that would prevent the team from removing nonprioritized deliverables from the project scope. At the end of the session, Karen and the team had defined a prioritized list of deliverables that they felt confident they could complete before the full application was scheduled to be released to investors. Because she was a key part of this process, Karen was ready to champion the new plan and communicate it to other stakeholders, including the sponsor. After all, Karen as product owner was fully invested in the success or failure of this project, so she was going to do everything possible to keep it on track.

At the end of the project journey, the project delivered the prioritized scope as promised. The application was launched on schedule, and it appeared the team performed a magic trick in the process. Rick was instrumental in this success. Behind the scenes, the reality was that the technique the team used to launch the application on schedule did not produce all the components originally included in the project. This meant that after the application launched, components were evaluated and prioritized according to the value they would produce for Blue Run. Because the project was not waiting on the completion of the evaluation process, Blue Run was able to do a more thorough job of gathering feedback from investors to identify what additional features and functions would be beneficial and compare those to other similar products in the marketplace. Investors now using the new application were able to be more specific about what they would need or want for additional functionality.

Long term, this approach does have risks that stakeholder satisfaction fades to become an illusion. After the elation of project success fades, like all magic, the amazement and awe of the magic feat may wear off when other real-world pressures quickly make the new application obsolete or ineffective due to the limited functionality that was produced to launch the application on schedule. Other priorities may push back any further development of the new application after the release.

5.6 Seeing Is Believing

How do you get people to believe the impossible is possible? Simple: Make it happen. There are plenty of articles that talk about the fallacies of delivering projects to complete scope, on budget, and on schedule. Do not let the challenges hinder your success. Remember, the real value you provide the organization is the ability to look beyond the barriers of what you cannot do and believe in your ability to create the magic that will amaze and delight your stakeholders. To produce this type of magic, you will need to engage a team of experts to work behind the scenes and sometimes on stage. These experts will help you prepare the stage and the tools that deliver the expectations to the audience—your stakeholders. To succeed, you must understand and appreciate that without a capable, cohesive team, the magic will not materialize.

Here is the biggest surprise of the show: The value produced through the magic is real. The magic is produced through your skills in leadership, critical thinking, communication, coaching, mentoring, teaching, and, of course, your expertise in the practice of project management to wield the forces of the Iron Triangle that represent schedule, scope, and resources to deliver measurable value and positive stakeholder experiences despite constraints that seem impossible to overcome. The illusion is the wand. It is simply the belief that a project tool or practice will successfully manage the forces of the Iron Triangle without true project leadership. Lose your reliance on the wand. Develop the real magic.

Chapter 6

Teleporting into the Future

© Jim Kangas

What would we find if we could teleport ourselves 20 years into the future? Would the world be without professional project managers? Would project management as a career choice be valued as a profession that is critical for any growth-oriented organization? Part of the answer we seek to these questions lies in how the world has seen project management in the past and how it sees project management today. To find the answer, let us first look to the past.

6.1 Project Management of the Past

Project managers of the past relied primarily on scheduling software and similar tools. Scheduling software was expensive and often restricted from use by staff who were not project managers. The team and the project relied on the project manager's ability to use these tools to forecast project complete dates, track expenses, anticipate the resources and the expertise needed, and manage the critical tasks. Critical tasks were part of the critical path in the project schedule and represented tasks that, at a minimum, were required to be completed to produce the project deliverables. Project value based on the work that was completed at any point in time during the life cycle of a project was also information the organization used to assess the health of a project.

In days past, it was typically the project manager who provided the data necessary to assess the current value of the project. The current project value at a given point in time is a factor known as net present value (NPV). NPV was one of many key factors in a decision to continue or kill the project. The project manager delegated tasks to team members, calculating the hours, days, weeks, and months that all team members needed to complete their assigned tasks. The calculated estimate provided a baseline at the start of the project that the project manager then used throughout the project to measure team and project performance. Actual performance against baseline performance was often used as an indicator to help the project manager determine at any point in time what the actual complete date would be for the project. Project managers labored over spreadsheets and reports to track, evaluate, and reevaluate scope, schedule, and cost performance so that, at any point in time, the project manager could provide an estimate of resources needed to complete the project and a best guess on when the project would complete.

6.1.1 Relevant for a Moment in Time

During this time, project managers also used other factors to calculate the project's value. All these factors had one thing in common: They predicted project value based on the projected investments of time and resource that would be required. As the pace of business increased and change became a constant in day-to-day operations, information collected based on a specific point in time became obsolete before the organization could act on the information. As a result, the tools and processes that produced linear, static information became less effective, falling short of keeping up to date with actual project performance and the pace of the business. In addition, the act of producing static information was a resource drain and a cost to the organization. Despite the inaccuracy of the process to predict accurate resource needs and project complete dates, annual planning at the organizational level and predictions on resource investment remained a baseline requirement so executives could prepare and fund projects throughout the organization. It was a linear model organized at the enterprise level that flowed throughout the organization. Projects and programs were not excluded from the linear planning model. Project management also relied on data that was gathered at a point in time, fixed, and often outdated from the point the data was communicated. This often meant the purpose of the project and the work that needed to be done to complete the project would change frequently throughout the project life cycle. In the past, the project manager was the main contact who ushered the project through all phases of the project life cycle—initiation, planning, executing, and closing—for the organization.

6.2 Project Management in the Present Day

Today, a wide array of project management tools is available to anyone who is inclined to use them. YouTube offers instant, on-demand instruction on almost every tool that is widely used. Scheduling software is available for free, or for a low one-time fee, from the Cloud. Project managers and even team members may use something as simple as a giant poster board and Post-it® notes to track task progress on projects. Project managers are still needed to monitor and control expenses and resources on projects. For every project that has human resources dedicated to working only on the project with teams that are co-located, day-to-day monitoring of resource allocation can become a wasted effort when the resource time is fixed, not variable, and based on a linear project plan.

6.2.1 It's About People

Project managers are divided and sometimes confused when it comes to agreeing on the best method to manage a project. Consultants seeking an opportunity to help the industry move forward are designing alternative methods of managing projects and portfolios. All the methods profess to be the answer to project and portfolio management, predictability, and performance. Project Management Offices (PMOs) in organizations are struggling to provide consistent returns from investments in project efforts. The environment is uncertain, and project performance remains unpredictable. Too many organizations struggle to understand the value of project management, while project management struggles to demonstrate value.

There is good news. Communication is still a primary need on every project. Although, depending on the size of the project, project communication can be completed by any team member because special software is no longer required, depending on the complexity of the project, special scheduling software such as Microsoft® Project Professional may still be needed. As time marches on, project management becomes less about the tools and software and more about people management and leadership skills.

Make no mistake, the project manager's knowledge of skills, tools, and behaviors is quickly evolving as one of the key factors needed for success in project management. The ability of a project manager to critically analyze and assess the right mix of behaviors, tools, and methodologies at the appropriate level of rigor allows the project manager to create the best possible project and stakeholder outcomes and experience. This ability is one of the key value propositions from project management to the organization. Project managers need to stay relevant in today's business environment and keep up with the requirements for success which have evolved from days past. Now, experts in many professions throughout the business world also manage projects for their respective organizations.

6.2.2 Adaptive Project Managers

Project management is a profession that is not just reserved for the career project manager. Individuals who are experts in a specific discipline in the organization while also managing projects are called adaptive project managers. Adaptive project managers have the same access to all the tools, processes, and training that are available to any career project management professional.

Translating the current project management environment back to the magician and the wand of Chapter 5, essentially, today there is a belief that anyone can hold a magic wand and create the illusion of magic on a project, because too often project value is experienced more often as an illusion than reality. Given this environment, and the predicted shortage of project management professionals by 2027, as reported by the Project Management Institute (2017), is there a risk that the career project manager will not exist in the future? After all, why should any organization invest in a PMO and a team of project management professionals if the magic they produce on projects can be duplicated by other individuals in the business?

This question highlights the single greatest threat to professionals in project management today. The answer may determine if a career in project management will be a viable profession in the future.

6.3 Project Management in the Future

If today's YouTube is any indication of the future for project management, professionals throughout the business world will be equipped with a do-it-yourself project management kit. The kit will be available on demand. It will include software, templates, self-help videos, and a variety of on-demand products and literature designed to help professionals in the fields of engineering, software development, marketing, or any number of other occupations throughout the organization, to manage any size of project while also completing the requirements of the job in their field of engineering, software development, marketing, etc. This practice started in the past with the advent of adaptive project managers, and it's growing exponentially.

In the past, using adaptive project managers was not considered best practice. Over time, it quickly became a popular practice because it provided organizations with the opportunity to merge two professions into one business professional. In the future, the appeal of using adaptive project managers will probably spread like wildfire, despite reams of brain and behavioral science that provide solid data demonstrating the poor results earned when one employee is used to fill multiple roles on a single effort. Organizations are often too busy to measure project returns or success factors once the project deliverables are completed. In the absence of performance measures for the project and the team, the opportunity is diminished to learn from mistakes, or identify and optimize successful practice and behaviors so teams and organizations can identify and adhere to a practice that enhances project returns. As the practice of using adaptive project managers continues into the future, project timelines will balloon, the completion of deliverables will be delayed, and project resource constraints will continue to be the number-one complaint that negatively affects project team performance.

As we look into our crystal ball to examine the future, we see that project management professionals who are dedicated to a career in project management and continue to use the behaviors, methods, and tools of old will no longer be valued in many organizations. With too few project management professionals (PMPs) in organizations and tools that do not keep pace with the needs of the business, the PMO may be practically nonexistent in the future. Organizations, with little hope of hitting project targets, will become complacent about project deadlines.

This view of the future might sound like an indication that the PMP is predestined for extinction, yet this version of the future does not need to be our destiny as a profession. After all, the future has not yet arrived. If you believe the profession of project management has the power to lead the way into a future that will ensure the value of the project management professional in organizations across the globe and sustain its value well into the future, keep reading.

6.4 The Future Is in Our Hands

Part of the path to understand how project managers forge the way into a bright, sustainable future lies in this book's previous chapters. In these chapters, we have looked at a variety of stories that represent common experiences in the life of a project manager. In each story, the project manager's ability to use emotional intelligence (EI), to adopt and exhibit leadership behaviors, is a determining factor that defines the outcome for the team, the organization, and the project. EI represents soft skills that will be in greater demand, along with other core social skills, than many of the technical skills we require from today's workforce (World Economic Forum, 2017).

6.5 Leadership

To determine whether all individuals have the ability to exhibit the necessary leadership behaviors outlined in Chapter 3, we need to tackle the age-old question, "Can leadership be taught, or are we born with leadership capabilities?" Dr. Guy Winch (2015) says that leadership qualities are only 30 percent genetic. This means most of our qualities are developed through our own efforts to adopt behaviors and skills that can make us successful leaders.

Dr. Winch brings us good news because his article reinforces the idea that we have the power to design our own future and create new opportunities along the way. Our power to design our own future lies in our ability to apply the knowledge and discipline to adopt new leadership behaviors that result in measurable positive results while keeping our commitment to change old unproductive behaviors.

6.6 The Next Step

The next step toward the future is simply to begin to lay the groundwork that sustainably assures our value is understood and recognized in our organizations, our communities, and by our colleagues. If we can complete this step, what might a day in the life of a PMP look like in the future? Let us take another look into the crystal ball.

6.7 The First Call of the Day

"Beth, can I talk to you for a moment?"

Beth reads the instant message on her computer, puts her headset on, and gives Kim a call. "Hi Kim, what can I do for you?" said Beth.

"I'm trying to set up this project schedule for our new product line, so we can track the projected launch dates for every new product we produce this year at Monkey Crackers," explains Kim.

Beth states reassuringly, "OK, I think I can help you with that, Kim. First, I'd like to ask you a few questions that will help me better understand what you need. Keep in mind there are several types of tools that can assist you in tracking dates and tasks, so going through these questions will help us both understand which tool is the right tool for you."

"That sounds reasonable, Beth," says Kim.

Beth continues, "Do you need to forecast projected launch dates for these products, or are you given fixed dates that dictate when you must launch?"

Kim carefully ponders the question, then replies, "We need both, really. Sometimes we have a fixed date that the director of sales gives us. Usually this means an obligation has already been made to the customer to ship the product by a specific date. If we miss the date, there may be legal and financial consequences. On the other hand, sometimes we can tell the director the target date we anticipate launching a new product. In that case, the date reflects the actual time we estimate it will take to develop, manufacture, and ship the product. In both cases, the original forecast date and the actual date need to be documented so we can predict any shortfalls in our ability to meet the shipping date. We need to complete the prediction for the shipping date in time for us to make adjustments that would help us make that date."

Beth considers Kim's needs, then replies, "Right, so I think I understand you need a tool that can help you forecast the anticipated launch date based on the estimates you and the team make to complete the tasks required to develop the production line for the new product and get it to market. You also need to track the project complete dates against the original projected dates, which is the baseline in your schedule."

"That's right," Kim says enthusiastically. "I need to be able to change or eliminate noncritical tasks, so I can provide scenarios to the team that might help us meet those target dates by only including what is absolutely critical to complete."

"Ah, so you need to create a critical path schedule?" says Beth.

"Yes, that's it," states Kim.

"Well, it sounds like you've picked the right tool for your purpose. You can set up a baseline for your start and finish dates and the work effort for each task in the schedule. You can also see the total hours per resource in the resource graph. Oh, and one more thing. You may want to show the total hours each team member needs to dedicate to the project to the team's supervisors, so they have a current understanding of the time each employee needs to spend on the project to launch the new product. The time each team member needs to spend on the project is important for supervisors to know before you start the work, so managers and supervisors can work to back-fill team members, off-load tasks, or remove team members from other efforts as needed, then they can dedicate the time required and launch the product in time. Remember, project management delivers the best value when used proactively, before the project work even starts, to provide information concerning the time each team member needs to dedicate to get the product delivered by the delivery date as planned. Does that make sense, Kim?" asks Beth.

Kim replies, "Oh, yes, I see what you mean." Then Beth says, "I have one more question for you. Why do you need a team to launch this new product? Isn't this something you normally do as a part of your job?"

"Actually, every time we do a new product that we've never done before, we kick off a project. The tools and processes really help us coordinate different teams, so we can create the new packaging, distribution channels, production lines, formulas, labeling, and all the variables that are new for us because we've never done this type of a product before," says Kim.

Kim thinks for a moment and then shares, "I think I know what I should do now. This helped a lot, Beth. You've given me some really good ideas. Thanks."

Beth feels energized when she gets off her first call of the day with Kim. As a PMP at Monkey Crackers, Beth helps the business every day as a coach and mentor for project management. Sometimes Beth receives a call from someone who is a project coordinator with a basic understanding of project management. At other times Beth receives a call from an adaptive project manager in the organization who is trying to juggle their regular job while managing a project. Acting as a coach and mentor to other people in the organization helps Beth use skills to support the efforts other people in the organization are trying to complete. Beth always feels good when she can say or do something to help. Serving the organization and the people she works with gives Beth purpose. Every time Beth completes a call like the one she just finished with Kim, Beth feels energized.

6.8 Teamwork

Beth glances at her computer screen and says to herself, "It's already 8:30 a.m. Wow, this day will go fast. I better get moving or the team will start the check-in without me." Beth goes quickly to the project room to lead the team's stand-up meeting. As a Certified Scrum Master (CSM), Beth helps the team complete the stand-up, a type of scrum meeting. This meeting is an efficient way for team members to run through what they completed yesterday, what they will do today, and note any concerns. The team has been working with Beth for two years now, and the group has achieved the performing stage in team productivity. Performing represents the optimal stage of team performance in Bruce Tuckman's team performance model (Smith, 2005). Over time, Beth's team has evolved through forming, storming, norming, and performing. In fact, the team is now so familiar with their roles and the processes they are using for the scrum meeting that the group of 10 often rolls through their updates in 10 minutes flat.

Throughout the stand-up, Beth listens closely to pick up clues on how different team members are feeling about their progress and the project, by honing into verbal, nonverbal, and behavioral cues from each individual in the group. Kelly, the team's database administrator, has stated more than once this week that while she has no barriers to her work, she has concerns regarding the integrity of the data the team is preparing to migrate from the old application into the application they are developing. Hearing this, Beth makes herself a note to follow up with Kelly later in the day to find out more specifics about Kelly's concerns. By meeting with Kelly, Beth will get a deeper understanding of Kelly's concerns, which will allow Beth to determine whether there are risks that need to be managed proactively, or there are other actions the team needs to complete.

Jane states she was not able to work on any project-related items because she had been asked to complete work on another project that was passed to her from the IT help desk queue. When Beth's project started two years ago, the team's supervisors were asked to protect the time for each team member needed on the project, so each member of the project team could be dedicated to the project work full-time. Beth knows that Jane should not be a resource for the help desk, because Jane is a key resource for this project. If the situation continues, it may affect Jane's ability to complete her work on time on one or both projects. It is also possible that Jane's supervisor is not aware of the situation or the impact on Jane's work. Beth makes a mental note to follow up with Jane's supervisor on the demands on Jane's time because, for the last two stand-ups, Jane has voiced her concern that she does not have the time to work on both projects. Beth knows that prompt follow-up on Kelly's and Jane's concerns is a key factor to preventing these concerns, which are essentially risks, from becoming issues on the project. Should these risks become issues, they could easily derail timelines, the quality of the deliverables the project is producing, and project success.

At this stage in the project, the team only needs Beth for some reporting, facilitation, and working to remove barriers that might prevent the team from completing the project deliverables. Beth smiles now as she remembers how different her role is today than it was at the beginning of the project. During the early initiation stage of the project, Beth dedicated about 80 percent of her time to getting the project framework, tools, roles, and processes set up with the team. Eight months later, Beth can really see her investment pay off. Once the team understood how to work together through the processes they created with Beth, coordinating handoffs and general workflow, things went smoothly on the project most of the time. When the team was ready to tackle a new segment of work, such as fixing software bugs and testing, the group gathered together with Beth's guidance to identify and make any changes to their approach that were needed. Beth worked with the team carefully to select, edit, and create specific tools that would meet the needs of the project while ensuring the best experience for stakeholders. The team's method of determining the process they would use to tackle a new task was not written or formal, because it was really about communication, application, and practice, with Beth serving the team as a coach, leader, and mentor throughout the process. Over time, without even realizing it, the team has been following the iterative cycles of Plan-Do-Check-Act (PDCA), to improve how they work together. After each stand-up, team members working on developing the application check in specifically to determine the next priority work item. The group also does a quick check to be sure each member has what they need to keep moving forward on the work. With a clear understanding of their roles and responsibilities, the process to determine what to do next and identify who will do what, is completed quickly.

The product owner is also a key factor for the team. As the representative from the business who is responsible for the product produced by the project, the product owner is fully engaged and works with the team consistently in the project room. As the team works, the product owner is very active in prioritizing what products the project produces first, based on the needs of the business. The process is also iterative and fluid for making decisions based on the business needs. This is necessary because the business priorities at Monkey Crackers are changing rapidly to help keep pace with competitive market demands. The process is a fluid dance to balance business needs and technical requirements, with everyone in the team striving to achieve a single goal: complete the full scope of the project.

Communication to coordinate handoffs between remote team members is handled instantly through a team blog that is reserved for the development work on the project. The instant message group blog serves as a tool to connect members virtually and engage them in specific IT-based tasks. As the daily conversation scrolls by on the computer screen, Beth can view the team conversations and handoffs anytime through the iterations of data validation, migrations, bug testing, and so on. As the developers and testers communicate and pass work virtually from one to the other, the blog captures the activity so other stakeholders can check in to view the progress too.

With the project team heavily engaged in their work now, it is time for Beth to head to the soundproof conference room where she will work for the next two hours of the morning.

6.9 The Recording Session

Beth opens her laptop to set up the recording software and begin the session. For the last few weeks, Beth has been working on recording a series of YouTube-style videos designed to help Monkey Crackers' staff understand how, why, and when to use tools in the project management tool box. These tools are created for use by non-project managers, so they are designed to be simple and easy to use. The idea is to create consistency in the practices and tools that are used throughout the organization while allowing for flexibility and innovation, so the tools meet the specific needs of the effort. By creating short, three- to five-minute videos such as "How to Conduct a Project Kickoff Meeting," Beth customizes the tools based on the culture at Monkey Crackers while adhering to the best practice prescribed by the Project Management Institute. Each video explains how every tool can be applied with two levels of rigor. The level of rigor, light or heavy, is explained in the video, along with scenarios on how the tools might be used. The videos are accessed through the company's intranet, so employees can freely view the information when they need it. Videos and templates eliminate the need for the organization to pay for external training or hire consultants, saving the company money while ensuring consistency in the practices and tools that are applied across all projects.

Beth knows that the viewers of the videos she is creating will likely miss the benefit of knowledge and experience about the behaviors and soft skills that are necessary to support successful management of projects. It is difficult to define the best practice and behaviors that will cover all the variables created by personalities, environments, and other factors that are unique to each project. This creates a risk that the projects that are not led by a project manager will fail.

Although Beth has reviewed the risks that are inherent in the approach to project management at Monkey Crackers, which include the cost of project failure, slipped project schedules, poor product quality, and team dysfunction, Monkey Crackers executives are willing to accept the risks based on their belief that eliminating the PMO, reducing the utilization of professional project managers, and using Beth to develop adaptive project coordinators in the workforce will reduce resource needs and improve project performance. To help mitigate the risks that can ultimately trigger project failure, Beth and the organization strive to empower the workforce to develop soft skills which will consistently improve project performance and outcomes. To close knowledge gaps, Beth is also routinely called on to set up projects with the appropriate tools, processes, and frameworks. Once the projects are set up, the team is engaged, and the project kickoff meeting is complete, Beth hands off the now-established project to other employees in the business to manage. Because processes are standardized, the tools can be adapted to the appropriate level of rigor based on the needs of the project, enabling employees of the organization to coordinate projects and earn positive returns. Over time, Beth's direct contributions have helped her build a reputation as the trusted expert and a project professional at Monkey Crackers.

6.10 The Lab

At 10:30 a.m. it's time for Beth to head to the project lab. When she arrives at the lab, Beth sets up the training room with five tables that each have five chairs. The project lab is a two-hour open session. Employees can drop by to get Beth's advice, discuss stakeholder challenges, resource constraints, tools, processes, and talk about anything relating to a project they are working on, regardless of the role they serve in the organization. Depending on the need, Beth flips from mentor to coach and back to mentor in the lab sessions to help employees learn and grow as a direct result of the interaction. In the lab, Beth works to challenge and stretch employee skills, recommending other resources both within and outside the organization that can help employees grow in the areas of leadership, communication, conflict management, and more, regardless of their job title.

The first employee to drop by the lab is Roberto. Roberto needs some help to evaluate possibilities for an internal or external solution that might serve as the answer for a project he has been asked to work on. Beth works as a mentor and offers a quick walk-through of a couple of tools that she feels may help Roberto evaluate and score viable options that fit within the project scope. As she demonstrates the tools, Beth provides examples of the process Roberto might use to evaluate the solutions. Beth also walks through the basics of gathering requirements from stakeholders to be sure Roberto understands what functionality and features from a technical and nontechnical perspective the solution must include to meet the requirements of the project sponsor.

Later, Mary stops in at the lab and asks if she can listen in on the session with Roberto, because the conversation might be something that will help her, too. Roberto agrees, and Mary joins the conversation. Before the end of the session, Mary asks Roberto if she can meet with him to discuss some of the practices that he is using which might be effective for Mary to use on her project. Roberto agrees, and Mary immediately feels better, knowing that she has a new ally to discuss best practices she might use to become more effective in her efforts to manage the project.

After Roberto leaves, Mary asks Beth if she can speak to her confidentially about a stakeholder management issue. Beth gets up to close the training room door, posting the lab sign-in sheet on the outside of the door so anyone else who drops by can reserve time with Beth beginning in 60 minutes. Once Beth and Mary are alone in the training room, Mary shares her difficult stakeholder story while Beth serves as a coach to reflect the situation back to Mary. Then, Mary works to define her next steps, being thoughtful to anticipate possible outcomes. After 45 minutes, Mary feels confident she has designed an approach that will build trust and demonstrate her genuine interest in developing a solid relationship with the difficult stakeholder.

Mary leaves the lab session with Beth feeling encouraged and determined. As Beth opens the training room door, she realizes the two-hour session for the lab has already ended. With five minutes left, Beth looks at the sign-in sheet. There are four names on the sheet. She immediately opens her laptop and emails the four contacts to ask if they have an urgent need to make an appointment with her. If their need is not urgent, Beth suggests that the contacts drop in to next week's lab.

This drop-in process creates a risk that attendance to the labs will be inconsistent. For example, no one might attend the session next week, then four people might attend the session the following week. Even though the drop-in approach is imperfect for the lab sessions, it creates another avenue of support for professionals throughout the business who are also managing projects, without the need for outside professional training. Because the labs are also thought of as a gathering place, another valuable opportunity is created for lab participants to develop a project community within Monkey Crackers. Over time, Beth has come to know the lab participants, their projects, and their stakeholders. The knowledge Beth has gained strengthens her ability to be a valuable resource bringing like efforts together, connecting people with similar experiences, and helping each member of the project community support others.

6.11 When There Is No Project Management Office

Monkey Crackers dissolved their PMO about 10 years ago. As predicted a decade earlier, experienced project management professionals (PMPs) and program managers (PgMPs) had become too difficult to find. As senior professionals retired out of the workforce and the need for project and program managers increased, fewer in the workforce entered the profession of project management. This gap put pressure on the PMO model, which depended on a hierarchy of professional program and project managers to complete projects within the organization. When there were no skilled professionals to complete projects, work queues increased, and project deliverables were less frequent, often with reduced quality. Therefore, the organization dissolved the PMO and in its place implemented a self-service model based on an array of tools and processes. The model includes information on the best use of the tools and processes and was put in place so anyone in the organization could manage a project of low to medium complexity and rigor. The information includes videos which are stored on the company intranet.

6.12 Sharing Expertise Using Technology

There is a sharing center on Monkey Cracker's intranet where staff can write about their project experience using a tool or best practice, then post the tool and supporting documents so others can access the information as well. Beth knows this process is better than nothing, but it still falls short of the formal, well-orchestrated portfolio management that a PMO offers. Large IT projects are managed by the CIO and the IT department, while all other projects are managed by various business units within the organization. To manage the shortage of expertise at the PMP level, Beth's employer chose to use Beth across the organization as a coach, mentor, teacher, program, and project manager. The two PMPs, Beth and her colleague, are now centralized in IT. Beth's colleague is used primarily to manage large IT projects, while Beth's duties remain very diverse. The company has realized they can use an experienced project manager with skills to lead people, projects, and programs to coach and mentor others on the internal projects that support growth as needed to keep up with the pace of internal change that is required to keep Monkey Crackers relevant in the market. The model at the project level works because Beth's employer does not produce projects for other organizations or consumers. It works because the projects are designed to renew and build the internal technical and business infrastructure. The work is complex and there are plenty of projects to complete.

6.13 Learning and Teaching

It's 12:30 p.m. and time for Beth to grab a bite to eat before setting up to teach for the Project Management Certification Program. The Project Management Certification Program is an internal eight-month-long program at Monkey Crackers that provides an opportunity for staff who have never managed a project, and know little of project management, to lead a small project for their department for the first time while learning PMI best practice. Through a mix of instruction, work sessions, and group collaboration, students learn how to critically analyze the variables in their project to identify the best tools and processes that will develop the greatest opportunity for success. The level of rigor needed for the project that is being worked is constantly evaluated as well. In this program, many students have their first exposure to the project community that Beth supports within the organization. The exposure helps them develop in one another a project support group. The support group bonds over time so when the students complete the program they are comfortable to continue to learn and grow together. Through the process, they are creating yet another resource to support projects at Monkey Crackers. The program experience that students receive is designed to provide a low-risk, high-opportunity environment for learning and doing. Students are encouraged to take on a small department project to apply what they learn as they learn it in the program.

The immediate application of a new skill or technique can be quite unnerving for some of the students. Despite the case of nerves experienced by some students, everyone agrees that the experience they receive through the program pushes them to stretch their skills, so they can reach new heights in productivity and relationship management. Once students graduate from the program, they are welcome to participate in the open labs and continue their growth to move forward in their learning journey.

6.14 Circle Meetings

In today's class, students will learn how to conduct and facilitate a circle meeting. Circle meetings are one of Beth's favorite facilitation tools because they are an ancient form of collaboration that is still very effective today to create engagement opportunities that encourage equal collaboration and participation in groups. Circle talks, or meetings, were common in ancient cultures. The custom of the circle meeting often included a talking stick. The talking stick was a prop to indicate who in the circle had the floor to speak. Typically, the stick was passed, rather than pulled to participants around the circle, giving everyone the opportunity to speak on the question or topic presented by the circle leader (First Nations Pedagogy Online, 2009).

Today, circle meetings provide solid positive results when used for group engagement. One of the factors that makes this possible is the form of the circle, which removes the dynamic of a seat of power—otherwise known as the seat at the head of the table—from the group experience. Beth knows that when she does not sit at the head of a conference table with a group of people, she is indicating she is there to be an equal team member who will strive to collaborate, not dominate. It is also important that Beth's intention to be an equal member of the team is demonstrated by Beth's words. Most important, Beth's intentions must also be demonstrated by her actions. Any message from Beth is made real by her actions, which makes it genuine to the team. Beth's actions quickly build her credibility along the way throughout the organization. In response, the team will grow to trust and respect Beth. For Beth, the circle meeting is one method she can use to accomplish a high level of credibility with the team without using any positional power or the need to create hierarchies within the group, which would threaten open collaboration and team innovation. Beth avoids any action that might reduce collaboration because she is keenly aware that reducing opportunities for collaboration and innovation will limit the team's ability to perform well together. Beth knows that collaboration and innovation has long been identified as staples of strong performing project teams.

6.15 All Voices Are Equal

Speaking of the positive effects of removing positional power, Monkey Crackers does not have hierarchical job titles designed to indicate power on project teams. This practice ensures that the idea, rather than the title of the one who generated the idea, is what resonates with project teams at Monkey Crackers. Rather than focusing on titles, the company has focuses on areas of expertise on teams, which indicate individual areas of interest and contribution. Operating without titles allows staff on project teams to switch roles easily. For example, on one project, Beth may be responsible for monitoring and supporting Bob's performance in managing the project and the team, because she has the level of experience needed on that project. At the same time, Bob gains more experience in the area or discipline that is core to that specific project. On another project, the roles may switch, with Bob guiding and supporting Beth's performance, because his skills are stronger in the particular area needed for the project they are working. Switching roles based on expertise, not based on title or positional power, also reinforces a culture of collaboration that is idea, not title centered. At Monkey Crackers, ideas are pooled in an electronic information center that filters the information. Employees can view the idea and select it as a project that should be worked, provided the idea is aligned with the organization's corporate and market strategic goals. When the idea is selected, it is queued up for approvals based on funding and resource needs. Employees who championed the approved idea are given the option to work on the project team that brings the idea to fruition, provided they are available. Through these steps, Monkey Crackers has developed a culture that promotes fairness, engagement, employee satisfaction, and innovation. This process, the use of adaptive project managers and the absence of a PMO, is manageable for the organization because they have fewer than 500 employees and rely on their own capabilities to constantly innovate, learn, and grow their business.

As the group of 12 students shuffles in for the Project Management Certification Program, they gather around the center of the room, each taking a chair to place into the circle. Beth sits on one of the chairs that forms the circle with the group, talking stick in hand. She begins the session by telling a story. It is a serenity poem. Beth does this because she knows it is an important step to setting the tone of the meeting before it begins. Beth has observed rising stress levels in some of the individuals in the group, so she decided to use the technique to share a calming word to help the group clear their minds from the many distractions of the day and refocus the group before the circle discussion begins. The technique is also an effective way to dissolve tension, stress, and emotion, helping participants focus on the here and now during their time together. Prior to the meeting, Beth emailed the meeting topic to students, so they are prepared for the circle discussion on the topic, "The Greatest Challenge I Have Faced on a Project." After Beth finishes reciting the poem and asking the group to take a minute for reflection, the talking stick is pushed clockwise so each participant can share their response to the discussion topic. Other members of the group who were not speaking can respectfully pull the talking stick to offer words of support, experience, and perspective when the speaker is finished responding to the topic question. As the talking stick passes from one member in the circle to another, the group bonds through the experience, sharing their vulnerabilities, challenges, and successes in managing and coordinating projects.

After the meeting, the students break into small teams of two. The group forms teams to mentor and support each other. Each student reviews and provides feedback to a partner on the project work he or she has done or is planning to complete. This is a terrific way for the students to share project experiences that they may be unwilling to share in the larger group, because they are able to share on a more personal level. Every participant understands that all discussions are held in confidence. During the session, a lot of learning occurs at a group and interpersonal level. Beth is inspired by the growth and awareness occurring all around her today. It has been a good day, and the day is not over yet.

6.16 Managing Programs and Projects

For the last two hours of the day, Beth is scheduled to meet with subject-matter experts (SMEs) from two departments who have received approval and funding to develop a program to help them develop automated tools and processes to enable the two departments to work together seamlessly on customer requests.

The program will be a large, expansive effort, because much of the information and data that is needed to respond to customer requests is stored on a variety of spreadsheets that are dispersed throughout the two departments. Redundant data, duplicative work, and conflicting information has resulted in customer confusion and dissatisfaction for years. Monkey Crackers has identified the program as a strategic initiative for the organization that is needed to create new growth and opportunities in the market.

The tangled web of spreadsheets is a direct result of growth at Monkey Crackers, which has outpaced the ability for the organization to develop sustainable systems and tools. Now, to remain competitive and grow the business in today's market, developing sustainable tools and systems is a critical priority for the organization. Beth is an immense help in this process, because her responsibilities do not end at the PMP level. Beth obtained her Program Manager (PgMP) certification and training through PMI as well. Through her PMI training and experience, Beth can provide business leaders with the appropriate support to develop and define the goal, objectives, and overall strategy they are seeking to achieve through a project program.

6.17 Aligning Expectations

As Beth works to understand the program goals between the two departments, she uses a series of interview questions to help the department SMEs describe what they envision the picture of success will look like when the program completes. As the discussion continues, Beth is also able to evaluate whether the SMEs, who will be defined as sponsors for the program, have a shared vision and goal. Beth does this because she knows it is critical to get sponsors on the same page. If the sponsors are not aligned in their vision of what is success for the program, Beth cannot move forward to engage the rest of the team and conduct the kickoff meeting. If the team is engaged before the sponsors solidify a single, unified vision, it will be impossible to communicate clearly what the team will be working in the kickoff and throughout the program. After an hour of discussion, Beth is assured the sponsors share a common vision.

6.18 Create the Roadmap, Define the Benefits

The next step is to begin working with the sponsors to define the program benefits and a roadmap that will describe how the projects within the program will deliver the capabilities and outcomes needed to realize the program benefits. Because the program is still being defined, Beth identifies the benefits and roadmap components at a very high level. At the end of the meeting, Beth and the sponsors have firmly established enough scope and success measures to confidently plan the remaining components of the program that are needed to schedule the program kickoff. As program manager, Beth will be responsible for ensuring that the benefits are realized as anticipated. She will also ensure that the program roadmap guides the project teams to complete the deliverables as planned. Beth will help the project teams for the program solidify their roles, responsibilities, frameworks, tools, and processes during the first six weeks after the program kickoff. At the end of the six weeks, Beth's role will move from strategic manager and tactical planner to mentor and coach for the project teams and their project managers. This will allow the managers to stretch their skills and completely own their roles with their teams while decreasing the time Beth needs to spend on the program.

The project managers who are chosen to manage this program are also enrolled in the Project Management Certification Program. Through the program, Beth can guide them through the project life cycle while they learn along with the rest of the students. This model is another example of how Monkey Crackers uses Beth's skills as a teacher and mentor to maximize resources and expertise across the organization in a way that promotes project efficiencies.

Over the next few months, Beth leads the teams to complete their implementation roadmaps. She works regularly with leadership to secure solid communication between the program, the respective projects, and decision makers throughout the organization to secure and manage project resources. She also links other efforts going on within the organization through the Project Management Certification Program and the active projects that current students and former graduates are working. In the absence of a formal PMO, Beth develops the tools and processes the organization uses.

6.19 Does Socialized Project Management Work?

The project management model at Monkey Crackers has succeeded in socializing project management best practice throughout the organization, yet the model falls short of centralizing project management in a way that provides a strong, continuous flow from the top of the organization through delivery on projects, because departments throughout the organization often fund, initiate, and attempt to manage projects without the benefit of coaching, tools, or project management best practice. When this happens and projects derail, it frequently creates a ripple effect that affects all the efforts Beth is working on, because Beth is called on to mitigate the damage and get the derailed projects back on track.

6.20 The Real Cost

Monkey Crackers does not measure the costs that affect the organization because there is no coordinated, holistic initiation or management of projects and programs from a central area of expertise. The projects initiated are largely for internal development, so there appears to be no cost to customer satisfaction. Yet every time the ripple effect occurs, Beth is intensely aware of the cost. The cost manifests in project delays which almost always increase project costs and reduce the estimated returns on project investments. Over time, this puts strain on all the resources throughout the organization. Often, delays on one project affect the resources needed on other projects, creating the potential of a ripple effect that can create delays and increase costs throughout many efforts in the organization for an extended period of time.

Because there is no centralized management of projects, resources, and schedules, the true cost of failure on a single project is impossible to estimate and manage. When project failure occurs without centralized project management or support, the organization does not have the ability to reduce or prevent one failure from affecting all efforts in the organization. This means that even though Beth's efforts are earning returns on a team and individual level at the organization, Monkey Crackers is exposed to the inevitability of project failures which are impossible to measure and drain resources. Monkey Crackers might be able to sustain the loss while the projects and the organization remain small. If the organization grows and the capital investment in projects increases, Monkey Crackers may no longer be able to absorb losses caused by project failures. In summary, the project management model that Monkey Crackers has chosen to use jeopardizes the ability of the business to expand sustainably, efficiently, and to support the business and technical environment in a way that allows the organization to keep pace with the competition.

While the project environment at Monkey Crackers falls short of an ideal project environment, the organization has successfully created innovative solutions to compensate for the shortage of professional project managers in the workforce and internal resource constraints while still promoting the development of tools and process based on PMI best practice that are specifically suited for the organization's needs. It is a world that was formed from necessity. The path Monkey Crackers has chosen began long ago. The path is not the answer, and it is not sustainable. The path will lead the organization to a dead end when the model is no longer sustainable. The answer to Monkey Cracker's need to grow sustainably and compete efficiently in the marketplace lies in the ability of its leadership to recognize the essential value of a holistic, coordinated, portfolio, program, and project approach that will offer consistent positive returns from investments in project efforts, teams, and the deliverables they produce. The answer to a sustainable project approach for the organization is the PMO.

6.21 The Future Is Now

Beth's story of project management in the future has already begun. Market-driven practices of self-serve delivery for software, training, tools, and practices giving any user access to technology in the workplace is now as easy as a click on a website. Organizations can have access to powerful technological tools and pay per month for the privilege. As a result, projects to implement technology that supports growth and expansion within the organization are generally easier to complete and less robust than in the past. However, these projects are more frequent, not less frequent, due to the accessibility and lower cost of technology. PMOs that refuse to innovate and update their practices are viewed as outdated institutions that slow, not speed, the organization's ability to keep up with global change. Project managers who believe that holding on to their knowledge of project management best practice, like the magician who holds onto his wand, choosing not to teach, mentor, or coach others in the organization, are outdated dictators who lack the leadership capabilities required to motivate, serve, and support team and project success. Organizations that do not measure the impact produced by project deliverables and the stakeholder project experience lose perspective on which practices and behaviors lead to success or failure. As a result, unsuccessful practices continue to be used, reinforcing a belief that project management is an outdated industry that slows success and innovation.

6.22 Design a New Future

The picture of the future is not all gloomy. We still have the power to rewrite Beth's story to one that tells of a robust, productive project management community that is fully supported in organizations through the development of healthy PMOs and the consistent application of best practice. While project management is failing to adapt to the needs of its customers in some organizations and communities, innovative project management is growing, supporting rapid change and adapting for success at the same time. The key factor that makes this possible is often overlooked, yet that key factor is the one common denominator across all the different models and methods that are rapidly evolving in project management today. That key factor is the project manager who leads from a position of service to the organization and the team and does not lead from a position of power, regardless of the project manager's job title or position in the organization. This is the type of project management professional who does not stand on a pedestal directing the team and the work on the project; this is the type of project management professional who serves as the pedestal to support the team, the project work, and the organization. Project management professionals around the globe are embracing leadership and answering the call to act. These individuals are serving in organizations and communities around the world as the pedestal that supports growth, innovation, and success.

6.23 Building Communities

The institution of project management is poised to create a sustainable future for the industry of project management. The PMI Educational Foundation, PMIEF, is providing project management knowledge and best practice to non-profits, supporting primary and secondary education institutions through project management excellence, and offering support and scholarships to youth to help prepare the next generation for the workforce. No-cost project resources are provided through PMIEF to help provide standardized tools and processes that aid schools and communities in their project efforts. PMIEF has been socializing project management for over 20 years. If the past 20 years is any indication of future growth, PMIEF will grow exponentially in the next 20 years.

Project managers who support PMI, PMIEF, the PMO Squad's Veterans Project Management Mentoring Program, or who participate in volunteer and mentorship programs in local chapters throughout the world, help to strengthen the profession and socialize best practice. PMOs who embrace leadership, answering the call to educate, mentor, and support project management within the organization, and socialize the practice of project management to empower others and deliver the value promised to the organization, strengthen an understanding and appreciation of the profession of project management. Project managers, PMI organizations, volunteer groups, and PMOs working together now can lay the path for a sustainable bright future for the profession of project management and the organizations we serve.

6.24 Design Your Own Future

Whether you are an executive in your organization, a member of a project team, or a seasoned project manager, stepping up today to lead and mentor change in your organization to become a valued project management resource as an advocate, advisor, educator, coach, and mentor, you will be a key part of creating a future that secures the success, growth, value, appreciation, and respect of project teams, project efforts, and the discipline of project management around the globe.

Chapter 7

The Chameleon

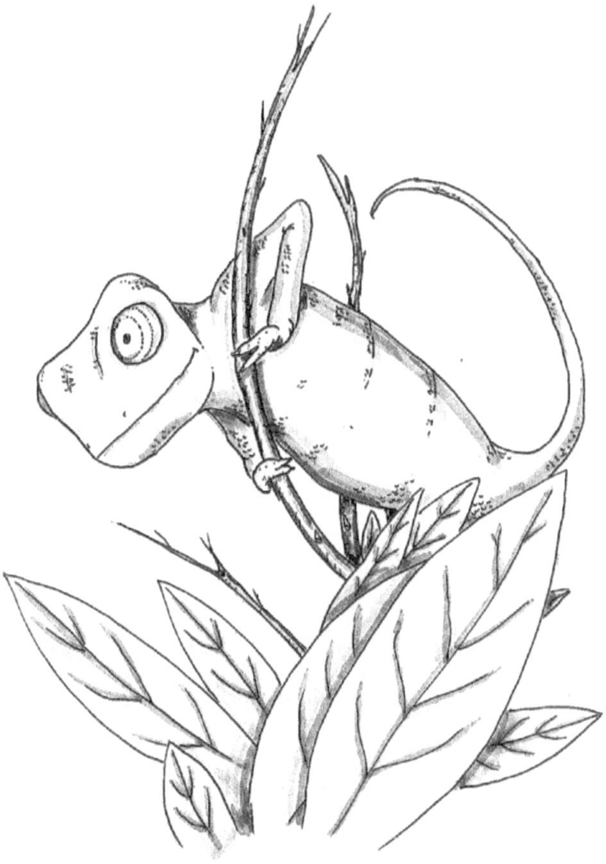

© Jim Kangas

In nature, the chameleon is a creature that adapts the color of its skin to the environment. The chameleon's ability to blend into its surroundings is an important capability that promotes the creature's survival into the future. Like the chameleon in nature, the savvy professional must have a similar skill set in today's ever-changing business environments to successfully adapt new behaviors and skills that are relevant to the environment. At the same time, the new behaviors and skills must be based on best practice to ensure value can be earned from adapting the new behaviors and skills.

7.1 What It Takes

The process for adapting new behaviors requires you to employ a few key skills, including a willingness to adapt, the capacity for change, and strong emotional intelligence (EI)—skills to identify what adaptions are needed to earn trust and acceptance in a new environment. Previous chapters in this book outlined why the skill to adapt is particularly important for project managers who want to succeed and the high EI that is required to adapt behaviors and skills in fast-paced, complex environments.

7.1.1 Communication and Culture

There is plenty of statistical support for the belief that timely and appropriate communication is 80 to 99 percent of success for any project. This information indicates that communication is an activity that needs to be carefully designed and executed throughout the project life cycle and the organization to ensure stakeholder impressions of project management performance match the reality of the performance. This means that if the project performs, the communication celebrates the practices that supported optimal performance and success. It also means that if the project fails, communication is transparent, supporting the communication of stories that can lead to development, improvement, and growth while building unity within the team and the organization. This is the type of communication that is built on honesty and transparency, which cultivates a culture of trust.

This type of culture recognizes that success is often achieved through failure. It's a culture that understands success is built on commitment, communication, and perseverance. A healthy team culture gives permission to fail so teams can learn, grow, and succeed. The journey depends on the right messaging, sent to the right people, in the right way, at the right time. How do we determine what defines "right" in this process, when it's different for every organization, stakeholder group, project, and person? The answer lies in the ability of the project manager to assess and exhibit the behaviors and communication needed for each stakeholder and each situation. The *PMBOK® Guide*, Sixth Edition (Project Management Institute, 2017), explains some of the many dimensions of communications, which include—and are not limited to—environment, culture, and stakeholder position in the organization.

Project managers need to be prepared to deal with people from all cultures. Project managers lead teams that have their own subcultures, organization, and industry culture. To do well across multiple environments and cultures, project managers must learn to adapt quickly and adjust their style to fit each environment and situation. Assimilating and adopting the customs and norms that are embraced within a stakeholder group and organization are important to becoming a trusted partner within a group. To get a clear idea of how this process works in our societies, let's examine the concept of matching style to stakeholder groups and activity in military teams.

No matter where you are in the world, military personnel have their own cultural norms and expectations. Some of these expectations are outlined in rules, procedures, and regulations, while others are nonverbal and relayed through behavioral reinforcements within the group.

7.1.2 Overcoming a Cultural Norm

The story of the Hacksaw Ridge Soldier is a fitting example of what can happen when cultures clash. The story of the Hacksaw Ridge Soldier is a historical account of Desmond Doss, who joined the U.S. Army to support World War II as a combat medic. The clash of cultures between the military and Doss occurred almost immediately when Desmond enlisted, because as a Seventh Day Adventist, Desmond did not believe in carrying a weapon or working on the Sabbath. Desmond's beliefs not only clashed with military culture, his beliefs also clashed with the expectations of his peers. As a result, Desmond was isolated and quickly became an outcast in his own troop while still in boot camp. Doss stood firm in his beliefs regardless of the cost. The cost included a loss of faith, trust, and respect from the other members of the troop.

Desmond's troop was deployed to fight the Japanese at a place called Hacksaw Ridge, on the island of Okinawa, in the spring of 1945. On one fateful day in battle, Desmond Doss rescued more than 75 men from what would have been certain death. Later the same year, Doss was awarded the Medal of Honor by President Harry Truman for his acts of heroism (Blair, 2016).

The story of Desmond Doss is an example of what can happen when an individual goes against the cultural norms of an organization and a group. It's an extraordinary story that demonstrates how breaking cultural norms can be accomplished in a way that earns trust and admiration. Even today, Doss is revered as an American hero. I wonder, though, how the story would have ended if Doss had not accomplished such extraordinary feats of bravery that day. If Doss had not saved any soldiers, the outcome might have been quite different. Doss's extraordinary courage brought extraordinary results, which diminished the importance of the cultural clash between the accepted military culture of carrying arms and Doss's religious beliefs that prohibited him from bearing arms. Why is this? Perhaps it's because courage, commitment, and valor are values held in the highest esteem in the military. Courage, commitment, and valor are values that have a predetermining factor of cultural fit and have the greatest importance in the military culture.

7.2 The Cost When Cultures Clash

Not every story that tells of a time when cultures clash has a happy ending. The massacre at Wounded Knee, South Dakota, in 1890 is one such account in history. In summary, the story tells of a Ghost Dance the Lakota tribe was conducting to call on the spirits to smite anyone who did not believe in the ways of the Indian. For years before the dance, great tension had been building between federal officials and the tribe. Promises had been broken, and times were hard for the members of the tribe. Fear and mistrust were silent participants in every interaction between the two cultures that widened the gap between understanding and compassion. Considering the divide between the cultures, it's no surprise that the U.S. government interpreted the Ghost Dance as an act of rebellion. It's not clear whether a lack of cultural understanding, or a motive to extinguish the tribe to clear the way for white settlers to mine for gold, thought to be prevalent in the Black Hills, was an underlying motive for the massacre that resulted from the events that occurred that day. When the sun set on that fateful day, U.S. soldiers had killed 150 men, women, and children of the Lakota tribe. This tragic story shows the devastating results that can unfold when two cultures collide.

During that same time and place, another story was unfolding in U.S. history. This was a story of a Sioux child who was sent to learn about and embrace the white culture. The account is equally telling about the impacts of clashes between cultures, yet the outcome was quite different.

Hakadah, who was later given the Sioux name Ohiyesa, meaning winner, was born into the Lakota Indian Nation during a time of great strife and suffering. As pressures mounted between the Sioux Nation and the white settlers, who were moving into the areas the Sioux claimed as their home, adapting to the white man's customs was viewed by many as a necessary survival technique. Ohiyesa's father, Tawakanhdeota, meaning the Face of Many Lightnings, was a Sioux who, after years of oppression, came to believe that adopting the white man's ways was the path to survival.

When Ohiyesa was 15, Tawakanhdeota changed his name to Jacob Face of Many Lightnings, and sent Ohiyesa to a mission day school to become educated in the ways of the white man. Ohiyesa then adopted the Christian name Charles, and took his mother's maiden name to become Charles Eastman. Over the next few decades Charles worked hard to achieve a degree in medicine and eventually become a physician. His first appointment was for the U.S. Bureau of Indian Affairs. It was in service as a physician through the bureau that Charles came to be one of the first doctors on the scene of the massacre at Wounded Knee. The experience had a profound effect on Charles, who just three years later moved with his wife to St. Paul, Minnesota, and established his own medical practice. Charles went on to spend his life navigating two worlds and two cultures that often clashed violently, trying to spread knowledge and understanding between the two worlds of the native and the white man. While Charles could not stop the tragedy at Wounded Knee, he participated in the development of organizations in both white and Indian communities that promoted leadership, education, and cultural respect. Those organizations, such as the Boy Scouts of America, still live on today. You can imagine what a difficult journey this must have been for Charles. The reward for his efforts was a place in history and a notable positive impact on generations of people who belong to both native and white cultures (United States History, 2017).

7.3 Bridging the Cultural Divide

These historical accounts of Hacksaw Ridge and Wounded Knee demonstrate the impact of choosing to adapt or not to adapt to a different culture and embrace behaviors, communications, and expectations that come with closing the gap between cultural differences.

In today's workplace, cultural diversity is embraced and encouraged. Most organizations understand that the richer the diversity in the environment, the richer the innovation within groups and organizations will be, and thus, the stronger the ability of the organization will be to gain market share and global recognition. Today, bridging cultural gaps between people and groups is not only supported, it's expected. So, how can we bridge cultural gaps? The key lies in our ability to adapt our communication style and approach to what I call the language of the listener. Speak a message that resonates with the listener's priorities. Use words that trigger feelings of respect and positive emotions while refraining from using words that spawn anger and disrespect. This might sound easy, but it's not. History is full of stories that show us the effort, and often, the cost required to adapt to a different culture or group. The first and most difficult step to transition into a new culture or group is to make a choice whether to accept the existing cultural norms, or reject them, and whether to initiate the effort for others to embrace your beliefs and behaviors or not. The decision for each of us is unique, and for each of us the decision holds different consequences. In most cases, the choice to reject or accept a culture is not based on right or wrong; it's based on which side of the cultural fence you hang your hat. There is a clear connection between the outcome, the choice, and how to communicate with others, both verbally and nonverbally. To demonstrate, let me share with you a few personal examples.

7.4 Adapting and Comfort

In my travels around the world, I have been confronted with decisions to adapt to the customs and cultural norms in any given environment or resist, and every time I have learned a lesson. As a young college student studying in France, I traveled with my girlfriend Jill to other countries throughout Europe at every opportunity. Jill and I were eager to learn about other cultures and really experience how people lived in cultures that were unfamiliar to us. On our first trip to Salzburg, Austria, we decided to visit the neighborhood public bathhouse once we got settled in at the local hostel. Before we departed for the bathhouse that night, we were told by the hostel keeper that the custom in the bathhouse was to wear only your birthday suit when entering the sauna, steam room, or pools. Of course, this was very shocking for two young American girls to hear and even harder to comprehend. Why would anyone want to visit a bathhouse of strange men and women when you were not wearing anything else but your birthday suit? Looking back, I now regret that neither of us did research or reviewed tips on how to behave in an Austrian bathhouse. So, ignorant of the importance of cultural assimilation in the Austrian bathhouse, we quickly changed into our swimsuits when we arrived at the bathhouse. Wearing what we were accustomed to in this unfamiliar environment would make us more comfortable, or so we thought. Once we walked through the changing rooms into the area where there was access to the sauna, steam room, and pool, the other patrons in the bathhouse began to whisper, shake their heads, and point as they sent us a clear message through their disapproving faces. The unrest of the locals made us uncomfortable; still, we pressed on because we were determined to experience the Austrian bathhouse. As we moved from the sauna to the pool, every person we encountered became visibly more upset. Finally, one tiny elderly woman stopped in her tracks right in front of me. I'm tall, so her presence was not particularly intimidating. Then she positioned herself within a few inches of my chin, which was uncomfortably close for my humble Puritan upbringing to endure, and proceeded to yell at me in German, the predominant language in Austria, while shaking her finger and her head violently. At that moment, and to this day, her actions sent me a clear message, even though I did not understand the words she spoke. Through her body language and tone, Jill and I knew that she was angry that we chose to go against the custom of nudity in the bath house. Suddenly the decision to stay with what we were comfortable with and not adopt the customs of this new culture seemed like a bad choice. That very decision now made us extremely uncomfortable.

After the woman stopped screaming, she turned on her heal and stormed away from me. Jill and I glanced at each other and knew instantly that we had made a terrible mistake in refusing to adopt the cultural norms embedded in this new experience. We headed for the changing room to dress and go back to the hostel with the understanding that we sacrificed the opportunity to experience the bathhouse and the culture the moment we rejected a cultural norm that made us feel uncomfortable.

Through that choice, we may have even sent a signal to the other patrons in the bathhouse that our way was the right way and any other approach to participating in a bathhouse was not acceptable. Of course, the signal we mistakenly sent was interpreted as saying something more profound and significant than we ever intended. In fact, we did not intend to send any signal or message at all. We simply wanted to experience the bathhouse in a way that made it comfortable for us. Our focus was on what made us feel comfortable. We were not focused on how our actions would make others around us feel. We were not focused on what behaviors and customs we might accept so that our presence would be more comfortable for everyone else. That was our mistake. In this story, our mistake cost us a good bit of humiliation, surprise, and regret. In reflection, it was a relatively small price to pay for the benefit of learning a profound lesson.

In the competitive world of business today, the cost to rejecting cultural norms in an organization or group is much greater than the price Jill and I paid. The cost can be job loss or even the destruction of a career.

7.5 Cultural Norms for Survival

Cultural norms and expectations are more than just customs we adopt that help us fit in or feel comfortable; they are standards we live by in a community or organization. Throughout history, in cultures across the globe, customs and norms were developed to keep groups safe and maintain survival. Cultural norms evolved as we as a species evolved.

One survival strategy of early native cultures was to engage in infanticide, killing of infants, and senilicide, the killing of the elderly, when conditions were too harsh and resources too scarce in the environment to allow for individuals who were not able to contribute to the survival of the tribe (Indacochea, 2006). While this practice might seem extreme or appear to be an act of murder, to those cultures that engaged in this practice in ancient times, senilicide and infanticide were believed to be a necessity to ensure the survival of the tribe. It was viewed as a sacrifice of one to ensure the survival of many.

Throughout history, in various cultures around the world, we know that sacrifice of a single life, whether voluntary or involuntary, was believed to be key to survival and prosperity for many ancient cultures.

In both personal and professional environments, for those unfamiliar with a culture or environment, it can be difficult to assess the value or necessity of a cultural norm without a deeper understanding of the environment and the culture in which it lives. Defining and leading positive change quickly in any organization is seen today as the hallmark of a successful leader. To assess rapidly what norms to accept and practice, and what norms to challenge so the right changes can be made at the right time, is critical to successful leadership. How can you know when to adapt and which changes to drive? How can you know the best approach, and how can you foresee the outcome to align your behaviors for success?

7.6 Piecing Together Change

There are plenty of formulas that provide guides for implementing change successfully in any organization or group. The ADKAR formula, which is a focus on creating Awareness, Desire, Knowledge, Ability, and Reinforcement, was created by the founder of the Prosci model for change, Jeff Hiatt, and is one of the more popular formulas that prescribe the process and steps for implementing change. Before you run to the website or order the book, take the step to gain an understanding the capabilities and limitations for change in yourself and others. Earlier in the book, I discussed that while we do have the power to influence and inspire change in others, we are not able to make people change their own behavior or accept change. Our greatest power to motivate and implement lasting change is through our ability to adjust our style and approach in a way that creates the desire for change in others.

Of course, if your title is Chief Executive Officer, you are empowered to inspire change in the organization by implementing rules, dictates, and processes as requirements of job performance. Project managers often do not have the advantage of positional power in an organization and generally must work much harder to implement cultural and behavioral change. For them, the success or failure of inspiring sustainable change in others depends on a couple of key factors. The main factor is that the project manager must possess or develop a high level of emotional intelligence, EI. Using skills in EI, the project manager reads the body language, verbal, and nonverbal clues that the project team, sponsors, and other stakeholders exhibit, assessing their reactions and feelings about the events unfolding at any given moment. Are you ready to give this a try? Before you do, it's important to follow a few simple guidelines.

7.7 How to Piece Together the Communication Puzzle

The process to adjust your style in a way that other groups and individuals can hear your message starts with your ability to hear, understand, and interpret the verbal and nonverbal cues they send you. It's like putting together a puzzle.

Have you ever put together a complex puzzle? Interpreting signals in other humans requires a similar process. First, you examine the picture of the finished puzzle. This is about visualizing the goal. To set a path to complete the picture of success, success must first be visualized.

Next, group together the pieces that are similar. It's easier to find your way to matching the pieces by grouping together pieces that are similar in colors and backgrounds.

Finally, and this is the tough part, once you match the groups, find what I affectionately call the, "bridge pieces." The bridge pieces are those pieces that share colors from two or more groups. These are the most important pieces. To match them between the two groups, evaluate everything about the puzzle, including the colors and shapes of individual pieces. The pieces that have flat sides provide important clues too. Looking at each piece, evaluating the clues and matching them to how they might fit in the final product by viewing the whole picture of the puzzle is the best way to complete the puzzle and achieve success. Success takes time. The time you invest is relative to the size and complexity of the puzzle. It takes perseverance and commitment. Often what starts as a fun adventure can become an arduous, frustrating task. After all, how long can a person stand to have 20,000 pieces spread all over the kitchen table? There is risk associated with the length of time those pieces remain on the table too. It's likely that leaving the pieces unmatched for an extended period will eventually cause pieces to get lost or damaged, making it impossible to succeed and complete the final picture. If a piece gets lost or damaged, you can attempt to put together the remaining pieces and accept that the outcome won't quite achieve the original goal and leave the puzzle incomplete. Let's demonstrate this process as we put together our own communication puzzle in the workplace.

7.8 Ben's Story

Ben walked into the project room and greeted the other 10 members of the team. The group had been working together for 18 months, and everyone was feeling the pressure to complete the final pieces of the project so the new application they were building could be released to staff in 30 days, as scheduled. Emma, the product owner for the project, was new to her role. In fact, she was new to quite a lot of things. Emma was new to the organization, new to her role as a department manager, and new to projects. She was eager to succeed in her new roles and willing to work hard to impress her boss, the CEO. To do this, Emma often felt like she had to be Superwoman and juggle whatever was thrown at her. Yet, throughout the last 18 months, regardless of the pressure, Emma had not dropped the ball on the project once. On the project, Ben hardly noticed Emma flinch when something new was thrown at her. Emma was there every time to support and defend the team when needed. She was open to learning innovative approaches that would help the project succeed when time and resources ran short. Emma even worked with the team to eliminate tasks that were not critical to the functionality of the application, to ease the pressure when resource constraints threatened to slip the project schedule.

As the project manager for this effort, Ben really appreciated the flexibility and support Emma offered the team. Her smiling face was often one Ben welcomed every day as he greeted the team. Unfortunately, that was not the case today. As Ben received a warm welcome, the usual greeting from everyone on the team, he noticed that Emma remained at her station in the team room with her head down and eyes focused on the screen. Ben thought maybe Emma had had a difficult morning at home. After all, everyone can have an occasional bad morning with the demands of balancing a young family and a new job. So, Ben chose to shrug off Emma's odd behavior and move on with the team morning check-in. As the team went round-robin to share their status on project tasks, Emma stayed silent. Ben thought, "This is unusual for Emma, I wonder what's going on." Finally, when the group finished their updates, Emma spoke: "Now I want you all to prepare to do whatever it takes to go live as scheduled." She added, "There will be no excuses for missing the deadline. If we have to work weekends or miss Thanksgiving, that's what we will do to reach our goal."

The team fell silent. To Ben, looking at the faces around the room, it was clear that the group was stunned by what they had just heard Emma say. This was highly unusual behavior from Emma. The team was a very hard-working, accountable group of professionals, all experts in their field, and Emma often thanked them for their dedication. Today was different. She was different. Ben knew there was something deeper going on with Emma, and he knew he needed to find out what it was—fast.

After the check-in, Ben walked back to his laptop to send an invite to Emma for a meeting later that day. After a couple of minutes, Ben noticed Emma accepted the invite. "That's good," thought Ben. "Maybe I'll be able to get to the bottom of what's bugging Emma before this day is over, so I can mitigate the damage before it disrupts the efficiency of the team."

As the day went on, Ben noticed a sense of tension in the project room. It was a kind of tension he had not felt before. This tension was not something you could see or even hear. Everyone was polite and considerate to each other as usual, yet there was something in the air, something you could feel. For most of the day, Ben observed the interactions between team members and Emma. In the process, he picked up another clue. None of the team members was engaging with Emma. This was highly unusual, because frequently throughout the day the typical behavior of the team was to reach out to engage with Emma, so they could get her feedback and direction. Ben knew the team was working on tasks that required Emma to review or at least be involved at some level at some point that day, yet team members seemed almost to go out of their way to avoid Emma. Ben thought this new behavior from the team might be related to the comments Emma had made that morning. "I'm sure the team is not happy with her approach. They are used to exceeding project goals, so the warning from Emma that there will be no excuses for missing the deadline may have upset or even offended the team. Clearly, Emma's message is having the opposite effect, demotivating rather than motivating the team today. I've got to get to the bottom of this fast," thought Ben.

As Ben walked to the conference room for his meeting with Emma later that day, he reminded himself to take it slow. "Listen first, find out how she is feeling, and see what happens," he told himself. As they sat down, Ben choose a seat across from Emma to be sure he could make eye contact while maintaining enough space so both of them would remain comfortable if the conversation became intense. Once they were settled, Ben opened the conversation: "Thanks for taking time to meet with me today, Emma, on such short notice. I realized it has been about a week since our last check-in, so I thought today would be a good time for us to catch up. I'd like to know how you are feeling about the project and the team."

Ben's approach, asking specifically how Emma was feeling about the project and the team, had stayed consistent from the beginning of the project. Ben's use of the word "feeling" was deliberate. This was important because Ben knew that asking a question about how someone is feeling gives them permission to share feelings that might be hard to articulate in terms of facts, data, or statistics. It's a critical piece of the puzzle, because our feelings govern our behavior more than any other factor. The workplace is no exception. That's because fear, and trust or a lack of trust, are feelings. The emotions of fear and trust are invisible bandits that steal productivity away from a team. These emotions stifle transparency and block collaboration between team members, and the damage they do does not stop there. Fear and distrust can spark other negative feelings and reactions between team members too. In short, fear and a distrust are the silent destroyers of a cohesive, well-performing team. These feelings are also difficult to deal with because most of the time, if someone feels distrust or fear toward another person, they will often not share these feelings with the other person. This might be because it's really hard to articulate feelings like fear and distrust unless there is hard data, or evidence of an undisputable incident to reference that can easily be identified as the source of these feelings. Most of the time, fear and a lack of trust come from our gut instinct. Yet, how often have you seen "the gut check says" in a report, or meeting, to prove or disprove something in the workplace? Most of us know that the response, "I'm listening to my gut," does not often pass for a plausible explanation in business. As a result, people often choose to engage in avoidance or exclusionary behaviors when they fear or distrust something or someone rather than deal with the issue. Avoidance was the behavior Emma had exhibited earlier that day. It was a signal to Ben that Emma might be feeling anger or distrust. The puzzle was far from complete. Ben needed a lot more information before he could understand what was driving Emma's behavior.

Ben waited for Emma to respond, making sure to engage in eye contact and display a calm, patient, empathetic demeanor. Emma took a deep breath and began to share her story. As it turned out, there was quite a lot to tell. Many things were going on in Emma's life that at face value appeared to have little to do with the project. As Ben listened patiently, he began to see the puzzle pieces fall together. As Emma continued, Ben realized that everything in the story had something to do with the project because it was affecting Emma, who was a critical resource for the team. Emma was telling a story about life events and how those events were making her feel. She felt people were constantly interrupting her, and because of this she was feeling she was not respected or valued. The CEO was putting pressure on Emma to meet the project deadlines, which made her feel accountable for the entire team and the success or failure of the project. This was not her role, yet she was made to feel her performance evaluation would be affected if the project did not deliver as promised. Emma had also been asked to cover for a vacancy that had recently came up in her department, and she had no idea how she would manage this in addition to everything else on her plate. Finally, Emma looked at Ben, her eyes beginning to water, and said, "As if all that I'm dealing with were not enough to handle, last week when I was trying to encourage the team and we were all sharing the level of confidence we each have in our ability to meet the go-live deadline, I felt you had no confidence at all in me. It felt like you were not supportive of my efforts."

Ben was shocked by this. Emma was one of the best product owners he had ever worked with. He was surprised and almost a little hurt by her comment. So he said, "Emma, I'm so surprised you feel this way. I can assure you I truly trust, appreciate, and respect you. Can you share with me what I did to cause your reaction?"

Then Emma told him why. Hearing the story, Ben quickly realized that there were no facts to deal with. There were no inappropriate words or actions. There was only Emma's impression of Ben's face in that meeting last week, when she shared her confident estimate of 100 percent for the team's ability to go live as scheduled. At this point, Emma was so upset her voice was shaking.

Ben was trying to keep a compassionate and understanding demeanor, but he was sure the profound confusion he was feeling was written all over his face. He wanted to lash out at Emma to defend himself. After all, she was being ridiculous to make such wild assumptions based on a look that washed over his face last week. Apparently, this has bothered her since that moment. On top of that, the assumptions Emma was making did not come close to representing how Ben really felt. How could Emma accuse Ben of not supporting her? Ben had always been there for her, mentoring her, coaching her, and giving her genuine positive reinforcement. He even spent countless hours listening to her challenges about work and home, challenges that had nothing to do with the project, or so he thought. Then, Ben reminded himself that this moment was not about him. It was not about what he had done or not done. It was about securing the success of the team. That could not happen if Emma and Ben could not trust and respect each other. So Ben shrugged off his feelings and refocused on Emma, who was looking at him earnestly, trying to assess his reaction to her words. Ben could see it took courage for Emma to share her feelings, so his first step was to acknowledge her effort. Ben said, "Emma, thank you for having the courage to trust me with your feelings. I am sorry I've given you the impression that I didn't believe in you. Let me assure you that I have immense respect and belief in you. You are the best product owner I've ever worked with. Every time a new curve ball gets thrown your way, you persevere for the team with grace and a positive can-do attitude. I'm not sure what signals you were reading on my face that day. I can only say I know I have been affected by other changes outside of the project in my own department. Since the reorganization two months ago, I find myself reporting to a boss who doesn't understand project management. The Project Management Office has been dissolved, and half of the employees who were my colleagues have left the organization. I must admit, I try to put on a brave face for the team but maybe I'm failing at this. I'm sorry."

At that moment a wave of empathy washed over Emma's face. Her entire demeanor changed. Her tense, concerned face immediately relaxed and displayed compassion and understanding. Emma said, "Oh Ben, I had forgotten all the changes you are going through too. Of course, now this makes sense. I'm sorry I misread your signals, Ben."

Thirty minutes had now passed in the meeting. Ben could tell by the look on Emma's face that she was relaxed now. After waiting a few moments, Ben said, "Emma, I'm really glad we connected today. Can you make me a promise?"

"Sure, Ben, what can I do?"

Ben stated, "In the future, let's make sure we get together right away if we are picking up distress signals from each other. This is so important because it will help us continue to operate like the cohesive partnership I know we are in front of the team. Our ability to operate like a partnership is one of the critical factors that helps this project to succeed. When our behavior is off, I notice it really makes the team uneasy. Now that everyone is under so much pressure, emotions are running high, so it doesn't take much to derail all the trust and confidence we have developed together. I promise to do my part to listen, coach, and support you and everyone on the team. Your support, trust, and honesty make this all possible, Emma. We couldn't do it without you."

Ben smiled and Emma returned an appreciative nod and smile.

"Emma, can I offer an observation?" Ben asked sincerely.

"Sure, Ben."

"When I listened to your story of all the things you've been dealing with lately, I thought I heard you say the CEO holds you responsible for bringing in the project on time?"

"Yes, that's right. He did say that to me," Emma reflected.

"Did you expect to be held accountable for the project timelines and delivery dates before he made that comment to you?" Ben asked, concerned.

"Well, no. I understood that was your responsibility. I'm new to all this, so when my boss made that comment, I thought maybe I was wrong about what I was responsible for," Emma replied as if she was looking to Ben for confirmation.

Ben sat up, looked straight into Emma's eyes, and said, "You are not wrong, Emma. It's my job to be accountable for project performance, our deadlines, and the schedule. I feel like piling this responsibility on you is unfair. It must be putting extreme pressure on you with everything else you are juggling."

Emma smiled, and Ben noticed her eyes beginning to water again. Then Ben offered, "Emma, I think it's possible that the CEO is pretty new to project roles too. He simply might not understand who is responsible for what. Even though we have shared communication on this, many times. I'd like to schedule a meeting with you, me, our technical lead, and the CEO just to touch base, review where we are at on the project, and give him the opportunity to ask questions. What do you think?"

Emma's eyes brightened, and she said, "Yes, I think that's a great idea. I know you have tried to meet with him and he keeps canceling. I'll try to work with his administrative assistant to be sure he shows up."

"Great. Thanks, Emma. Is there anything else you want to cover before we wrap up for the day?" Ben asked.

"No, I think I'm good now. I'm so glad we met, Ben. Thanks for all you do."

Emma and Ben left the conference room and walked into the project room smiling.

Ben's story demonstrates the importance of putting together the puzzle by interpreting the chain of words, actions, behaviors, and signals. You may have noticed that Ben had to temper his responses and reactions to get his relationship with Emma back on track. This story does not demonstrate what Ben would do when he needs to change his style. In the future, Ben might have similar challenges like the one he just experienced with Emma, if he displays a facial expression that communicates a signal he does not intend. As Ben continues to work hard to build his reputation as a trusted expert in the organization, he might be fortunate enough to receive feedback on how his nonverbal signals are affecting those around him before those signals negatively impact how people respond.

By now you might be thinking this is not really a problem. After all, how frequently do people assess a mood, attitude, or reaction from other people based on their facial expression alone? The sad truth is that we all assess other people, their mood, and their demeanor within less than a second after seeing their face (Wargo, 2006).

7.9 What Does Your Face Say?

One of these facial expressions is known as the "resting bitch face." This is a state that refers to the facial expression that displays a type of negative emotion that is interpreted as disgust, boredom, or irritation (South Palomares and Young, 2017). Perhaps you have been asked what's wrong when someone has simply looked at your face, or perhaps you have glanced at someone else and thought, "Wow, they look irritated. I guess I'll talk to them later." The trick in managing your ability to send clear, accurate signals lies in the awareness of how your demeanor and facial expressions are interpreted and in the ability to alter your expressions when sending signals that inspire communication.

7.10 Nonverbal Cues and Trust

Awhile back, I was hired on as a senior program manager for an organization that was relatively new to project management. The Project Management Office (PMO) had been in place five years and the organization was still learning practices, vocabulary, and tools. I was hired to manage a large, multiyear program and work with the executive team to set up the program and produce the deliverables. As time went on, I continued to build my relationships with the executive team. Eventually, every executive came to trust me, except for one, the CFO. Time and time again, I noticed that Amit, the CFO, would delay making decisions in meetings. He would follow up every meeting with me and my project teams by asking for more data and documentation, even when all the data had been presented in the meeting. This behavior slowed project process and frustrated the team. Every time I met with Amit, he gave me very little feedback, positive or negative. I tried several different approaches to get feedback, yet nothing changed. I couldn't shake the feeling that Amit was not being genuine with me. Over time, I became less trusting of Amit and other members of the executive team. With other options exhausted, it was time to share my concerns with my boss, the director of the PMO. "Debbie, over the last few months, I feel like I've run out of options to find out what's really going on with Amit," I stated, concerned.

"What do you mean?" Debbie asked.

"Since I've started here, I've felt Amit doesn't really trust me, and I just can't figure it out," I replied. "The situation is starting to interfere with the team's productivity and our progress. I've got to figure this out."

Debbie looked at me and responded slowly, "Denise, if I give you some feedback that may help, can I trust you to take it in stride?"

I sat a little straighter in my chair and said, "Of course, Debbie, you know I only want to improve our efficiencies and opportunities for success. If the CFO doesn't trust me, I won't have much chance of earning success, will I?"

Debbie continued, "I think the problem might be that you smile too much."

I sat in my chair for a moment dumbfounded and uttered, "What?"

Debbie continued, "You see, Amit just doesn't trust people who smile a lot. You smile a lot and are super upbeat, which could reinforce Amit's impression that you are not trustworthy or genuine."

At that moment I knew I had some hard choices to make. I thanked Debbie for her candor and left her office perplexed. I knew my positive demeanor was good for the team. It created an environment that was encouraging and fun. Genuine positive attitudes can be contagious just like genuine smiles and laughter, and that was the case with me. I feared if I changed my demeanor whenever Amit was around, there was a risk that I could be sending mixed signals to the team. This might negatively impact the trust I had earned with the team and make me less approachable. Over the next several months I trained myself to temper my enthusiasm around Amit while remaining true to how I genuinely felt. I learned to speak less in meetings and empower other members of the team to present data and information on behalf of the team. This seemed to improve the situation and Amit's concerns, which lightened the burden on me and the team.

Although we finished the program and the deliverables as promised, looking back, I don't think Amit ever came to truly trust me. Maybe he felt this way after the first millisecond of seeing my face. Even though I could not completely change Amit's personal assessment of me, I could change my approach just enough to reduce the impact of Amit's feelings toward me, so it would not impact the project or the team. The trick was to become a chameleon, so I could change my skin while still remaining genuine and true to whom I really was. Just a few changes in my approach and the way I engaged the team, over time, made the difference we needed to remove Amit's doubts and earn his confidence.

My story is not one that tells of a glorious tale of success, because in the end, I don't think I changed Amit's feelings about me. The projects, teams, and their success are not about me, the project manager, are they? Project success is about the team and whatever we can do together to overcome the obstacles and barriers to succeed, and that is exactly what we did, together.

7.11 A Flexible Approach Is Required

People and environments are dynamic and fluid. In an ever-changing world, leadership in project management requires a flexible approach. Like the chameleon, to become a leader, the ability to adjust your approach to one that best resonates with your environment is a requirement for survival. The key is first to define how you can best serve the team, stakeholders, and the organization, by adapting the tools, processes, and your own behavior.

Chapter 8

The Thrill of the Ride

© Jim Kangas

Think of a time you were sharing an experience with a group of people and one person in the group did not want to be a part of the experience. What was it like? Typically, if just one person in the group was having a rotten time, those negative feelings quickly spread throughout the rest of the group, weighing down the mood with negative tones and resulting in a terrible experience for everyone. Now, think of a time you went on a roller coaster, or shared a new experience with a group and the ride was a thrill of a lifetime. What was that like? If given the opportunity, which type of experience would you want to repeat?

Most people desire a thrilling experience when they start a journey, or a new adventure, as opposed to a negative experience. On teams, a new experience can quickly become a thrilling ride as the group travels through the adventure together. As the tour guide for the journey, the project manager is a key factor that will determine the type of ride the team will experience during the journey.

Tap into the passion within each individual on your team and you unleash brand new possibilities that make what was impossible, possible. This kind of passion is the kind that changes our world. It is the kind of passion that makes the ride the thrill of a lifetime.

While exploring the road to tap into the formidable energy that passion fuels, we are reminded that in nature there is a positive and a negative, a yin and a yang, side of passion.

The first step in tapping into passion is to determine whether it stems from a positive or negative force. The next step is to manage the negative force to keep it from spreading and growing stronger, while nurturing the positive force.

8.1 The Spoiler Virus

Science tells us that feelings intensify when we share an experience with others. As a result, when we experience something with a group of people, we experience either intensified positive or negative emotions. These stronger emotions are exactly what we experience if just one person in the group is not motivated or excited about being included in the experience and expresses a negative attitude. The negative attitude flowing from one individual is a factor I call the "spoiler virus." The spoiler virus can ruin the joy of the experience for others as well as drain the productivity and energy from any group.

When I think of my life as a single mom, I remember experiencing and fighting the spoiler virus almost every time I did anything new with my two girls, Alexa, age 8, and Anna, age 10. I recall one particular family experience that demonstrates the struggle. As you read my story, imagine how behaviors and actions in the story parallel life within project work teams. The analogy between family and project teams experiencing something new together is rooted in many of the same emotions and personality traits. Behaviors in our work teams and in our family units are driven by the same basic needs, such as the need to be acknowledged, to be cared for, to feel safe, to feel empowered and supported, and to feel valued by contributing to a purpose greater than ourselves.

8.2 Creating the Thrill

Day 1

By the time the first day of our family vacation arrived, I had created, reviewed, and digested the plan for our summer vacation with my two children, Alexa and Anna. We all contributed to the plan, and it became final when every family member agreed to the plan. The goal was to get to a point where we each felt the plan was our own, by including particular events that we wanted to experience. The events were named and managed by the person who chose them, so we could each make significant contributions to our vacation experience. The process made it easier for me to get buy-in to the plan from Alexa, who didn't really want to go to the Louvre Museum and wanted to go on the boat ride on the Seine River that was scheduled immediately after our visit to the Louvre. The sequence of this plan inspired Alexa to look forward to the tour of the Louvre so she could experience what she really wanted, the boat ride on the Seine.

With weeks of planning and discussion behind us, the big day finally arrived. We put our backpacks in the car and headed to the train station. Transportation by train was a selection I made with the intent of adding even more adventure to our itinerary. Neither Alexa nor Anna could remember the last time they had been on a train, and it had been years since I had traveled by train. Besides, I thought we would travel by bus and plane for most of the trip, so this was our chance to try something different. I had little experience with timetables for passenger trains, so I worked in a 24-hour time buffer between the time we were scheduled to arrive in Chicago and the time of our departing flight from Chicago's O'Hare International Airport for the next stop on our itinerary—Paris, France.

We arrived at the train station about 45 minutes prior to departure to enjoy lunch at the station and get our bearings before embarking on the train. We were excited and brimming with anticipation for the wonders we would explore. As we enjoyed lunch, our spirts were high. With only 15 minutes until our train was due to depart, I paid for our lunch and we scurried to the platform to embark on the train. When we arrived at the platform, though, I quickly realized there was a problem: there was no train. In fact, there was not a single train in sight. Hmm, I thought, "Now what?" After waiting patiently for 30 minutes, the kids were getting anxious and the spirits in our group were beginning to plummet. Eventually, a voice on the station loudspeaker announced a two-hour delay and instructed passengers to stay close to the station. Because this was a new experience, we did not have any information with which to make a new plan, so we stayed at the station. Seven hours later, our train finally arrived at the station. By the time we pulled into the Chicago station several hours later, it was midnight. As we searched for a cab outside Union Station, tired and disoriented, we plunged into the chaos of passengers outside the station scrambling to grab a cab.

Finally, we arrived at our hotel safely, overcoming our first hurdle of the trip. The bad news from the day was that our first sense of a new experience on this adventure was not a positive one. I knew the negative experience on the first day meant I would have to work a bit harder to keep the positive vibes going with the girls for the next leg of the journey, the 14-hour plane ride in coach class to Paris.

Day 2

The tasks we needed to complete to get on the plane that morning fell mostly in my court. My job was to make sure nothing and no one was left behind, keep my composure, stay positive, and tackle every situation with grace. To keep my composure and exhibit grace in every situation presented a bit of a challenge for me because I also was feeling anxious while leading Alexa and Anna through this new experience. I cared about my children, for their safety and well-being, and very much wanted our experience to be filled with positive memories. I desperately did not want to create the kind of memories kids talk about that include an episode of "Mommy Monster." Despite my anxiety, I knew the burden to create a fabulous experience on this trip was not mine alone. Each member of the family understood that we were a team and all team members had their own responsibilities. It was made clear before we embarked on our adventure that fulfilling our individual responsibilities would be critical if we wanted to have a positive vacation experience. Even with all the preparation and sharing of responsibility, I could not help but feel responsible for everyone's happiness on this trip. After all, I was the Mom and leader of our group. While putting the responsibility for everyone's happiness during the trip on myself may have been unfair to me, it did give me extra incentive to find ways to make the grueling travel schedule fun.

All of us donned our backpacks before heading for the airport that morning. The train fiasco had robbed us of a full night's sleep, and I could feel the anxiety and exhaustion in the group. Although I was tired, I kept my attitude upbeat, reviewing with the girls what the arrival at the airport would probably entail, and the options we could choose from once we landed in Paris, being careful to let Alexa and Anna choose from the list of options I presented. When each child chose a different option and we didn't have the time to do both, we took turns.

Soon, it was time to board the plane. Looking back now, I'm not sure I managed the security line with the most composure amidst the flurry of passports, tickets, backpacks, shoes, forgotten water bottles, and additional security checks. Fortunately, at the end of the line, we got through the security check together to board the plane for Paris with our belongings and my mind fully intact.

During the flight, I worked with Alexa and Anna to prepare them for arrival and for Paris. When we had made our rough itinerary together, it allowed plenty of time to choose additional activities we might discover once we were in our new environment. While on the plane, we also made some agreements that we promised to follow while we were in Paris. The agreements were essentially a list of simple rules that Alexa and Anna created with me to help the group have a positive experience.

Agreements were a good method that helped us to set and clarify expectations with each other about how we would behave to keep each other safe and informed while still having fun. Some of our agreements helped us remember to have fun, such as "Eat ice cream once a day" and "Visit the bakery daily." We also created a few safety agreements, such as "Stick together and never leave the group without permission," and "If you are lost in a building, go back to the door that we used to enter the building and wait for the group." A few of the agreements I contributed were designed to help us immerse ourselves as much as possible in the French culture and environment. This meant we used only public transportation and spoke French as much as possible. To help immerse us in our new temporary environment, we picked a small family-run hotel in Paris to help us escape the American tourist scene. When everyone had had a chance to add their agreements, we reviewed and refined them until we all understood and agreed to them.

By the time we arrived at our destination 14 hours later, we were eager to find our hotel and settle into our new temporary home. Just 30 minutes after we arrived at the airport, following the agreement to use only public transportation turned out to be the first agreement I desperately wanted to break. I knew hiring a cab would be easier for me than taking the subway, yet as the guide, mentor, and leader for this trip, I understood that if I broke an agreement, Alexa and Anna would quickly follow suit. So, I quickly assessed the inconvenience of navigating the subway was worth keeping our agreements intact for the rest of the trip.

I had studied in Paris as a student, so I felt comfortable with the language and the city, yet it had been over 20 years since I had been to Paris, the City of

Light. I was a bit nervous because so much time had passed since I had visited Paris, and I knew much of the city would be different. Speaking and reading the language might be a struggle for me now. When we got off the plane, my language skills were put to the test, almost immediately, to navigate our way safely to the hotel. The task was a lot more challenging than I had anticipated, because our hotel was tucked in a small alley off a busy street about a mile from the subway stop where we disembarked. After studying the map and asking for help from pedestrians at almost every block, we finally arrived at our hotel.

That night we took time to review our accomplishments of the day, relax, talk about the lessons learned, and absorb the sights and sounds around us in this new, unfamiliar environment. Next, we talked about the plan for the next day. As we talked, I carefully assessed the enthusiasm level of each of the girls when taking about an activity that we might try. I knew that Alexa was not comfortable in crowds or with strangers, yet she loved to explore. Anna was enthralled with kingdoms, princes and princesses of any kind, and she never gave a second thought to being alone. I used Alexa's and Anna's interests to spark their enthusiasm for the next day's activities. I eagerly told the girls, "Tomorrow, we visit the Louvre, a museum that was once a castle of kings and queens. We will explore with a guide and then explore on our own to learn all about the royal court, the Egyptians, and many famous artists. Then, we will hop on a boat down the Seine River."

As we enjoyed our ice cream that evening, I saw the girls' eyes sparkle with anticipation for the adventures to come. For these were the adventures we had designed together.

Day 3

The next morning, we were brimming with energy and excitement, ready to start the day. As we enjoyed breakfast at the corner bakery, fulfilling one of our agreements, I reviewed the options and benefits for our public transportation with Alexa and Anna. The girls chose a double-decker bus to whisk us through the city to our destination, the Louvre Museum. When we arrived at the Louvre, it felt like we were stepping into a magic kingdom. Alexa and Anna were jumping for joy and smiling ear to ear, and so was I. We all chose a section of the Louvre that we wanted to learn about and share with each other. Anna chose to learn everything about Charles V, the first king to use the Louvre as a royal palace, Alexa chose to learn about the Egyptian sarcophagi that were displayed in the west wing, and I chose to learn about the many paintings of Napoleon scattered throughout the museum.

Hours later, our feet were sore from walking miles through the museum. Later in the afternoon, we came upon the royal quarters of King Charles V. It was a marvelous re-creation of the king's living quarters. Anna was almost floating with excitement. The king's crown jewels were in the center of the room, protected within a glass case. When Anna spotted the jewels, she flew immediately to the glass case. After gazing through the glass case at the jewels for almost 10 minutes, Alexa and I were ready to move on, but Anna was not ready to go to the next exhibit. So, we made a deal. We would move on to the Egyptian wing and then swing by the jewels again on our way back to the exit toward the front of the museum. To this, Anna reluctantly agreed. Less than 15 minutes later, I turned to ask Alexa what she thought of the display of Egyptian sarcophagi we were viewing, only to realize that Anna was nowhere in sight. Alexa's delight quickly turned to panic as she realized her sister was missing. I felt panic too. At the same time, I knew that as the leader I needed to keep my emotions under control. Now was the time for me to exercise logic and manage the careful dissection of events so I could quickly determine what happened to Anna. Time was our enemy in this scenario, and I knew it. I placed my bet that Anna had gone back to the royal chambers, so I quickly headed in that direction. I didn't think I could breathe until I turned the corner and saw Anna gazing through the glass at the jewels.

Anna had broken the first and most important agreement, to stick together. I gave Anna a big hug and expressed how grateful I was to find her safe. Then, we sat on a nearby bench so the three of us could review why the agreements were important to each of us and the risks of not following the agreements we had made together. Once we completed our discussion and tears were brushed away, I hugged Alexa and Anna. Then we got up from the bench to continue our adventures for the day together.

The Rest of the Story

Throughout our six-week trip, we experienced more bumps along the way. Each time we ran into a challenge, we calmly worked through it together, reaffirming our agreements or creating new ones as needed. Looking back, I think it was relatively easy for the girls and me to get through the challenges because of the process that we followed and because Alexa, Anna, and I knew that we each had the best intentions. Most important, the girls knew I cared for their well-being and happiness more than I cared for my own. To this day, we reflect on the adventure to France as a thrilling family experience. It was a thrill we will never forget.

8.3 Practices at Home and Work

At work, at home, and in our communities, people form teams to come together to accomplish a mission. Whether it's a church building committee, a school board committee, a family tour group, or a corporate project team, teams behave and respond similarly because they are formed by humans. As humans, we all have the same basic human needs, feelings, and fears.

Personalities and behaviors in work environments are often more opaque than personalities and behaviors in personal environments. Regardless of the environment you are in, people and emotions are the common denominator from which you must base your approach. If you want to succeed in developing teams, you need to recognize that the ability to create project success is about developing and promoting healthy relationships and team environments. This requires a few simple steps that were demonstrated in the story of my journey with Alexa and Anna. As you navigate your journey, remember you are the guide for the team. Understand and embrace this role. The role carries with it the responsibility of modeling healthy team relationships to develop team building, not breaking behaviors. To fulfill the responsibility successfully, the guide must wear many hats, including those of leader, mentor, coach, caretaker, and role model.

8.4 Involve and Empower at Every Opportunity

To involve and empower others, you will need to remember that the journey is never about you. The goal is to focus on helping others imagine their own success and the success of the team. In the process, you will demonstrate that you hold each individual on the team in high regard. As an equal member of the team, you have a vote on the direction, the approach, and the tools you and the team use along the way. Others must feel respected, appreciated, and valued by you, the leader. In turn, the trust and respect each team member feels for you will grow. The process of shared trust and respect will create a safety net within the team that will encourage individual growth, inspire innovation, and provide empowerment.

8.5 Make It Fun

This one is easy. Think of small inspirational ways to develop in others a desire to participate and be a part of the team. You will know you are accomplishing this when you hear comments from team members such as, "I never want to leave this team" or "I love this team room, it's such a happy place." This can really happen at work. The first step is to believe it's possible.

8.6 Take Time to Appreciate Individual Contributions

It all starts with your understanding that everyone on the team, regardless of their expertise, has gifts to contribute. If some team members are not contributing or their contribution is negative, that is a signal that you have some extra work to do to help those individuals tap into their gifts and become inspired to share those gifts. If you are confident you have done everything possible to help team members feel valued and empowered to contribute and find they are still resistant, consider whether you may have to recommend they leave the team to prevent the spoiler virus from spreading. Exercise caution here, and remember everyone has something valuable to contribute. While typically you can't make everyone willing participants in anything, as a leader you can tap into their passions to find the trigger that will inspire them, ignite their energy, and harness their enthusiasm to support the team. Before beginning the process to remove nonperformers from the team, take a hard look at your own performance first, to assure you have truly practiced leadership in service to support all the members of the team.

Support and reward the team. If you are reading this and think, "That's not my job. People are rewarded by their supervisors for their performance," think again. In your role as team leader, the team looks to you for direction, approval, correction, and support. Showing support and appreciation for the team is important because dedicated project teams may spend 36 hours or more together per week. According to an article published in *The Washington Post* in 2015, spending 36 hours together with a team per week is almost triple the time U.S. mothers spend with their kids per week, and on average, more than quadruple the time U.S. fathers spend with their kids per week (Schulte, 2015). The UK *Independent* notes a similar time constraint per week in families, reporting that families spend an average of only 49 minutes per day together (Amara, 2010).

In both family and work groups, the quality of time you spend together is up to you and every member of the group. If you want a positive experience, you need to invest in building positive relationships. After all, according to the time we spend with our work teams versus our families, we have about four times the opportunity to connect, build, and model positive relationships within our work teams. Based on the time we spend at work, if we choose not to invest in building positive relationships, we might be spending most of our time in negative relationships. Can you imagine spending 65 percent of your time in a negative, nonproductive environment?

8.7 Walk the Talk

Have you ever been to a dance club and wanted more than anything to jump up on the dance floor to your favorite song, and found yourself anchored to your seat out of fear that you would be dancing alone if you didn't wait for others to get on the floor first? Most people do not want to dance alone. The risk is that no one else will dance so you sit it out and miss the opportunity. Is missing the dance a better option than taking the chance that you will dance alone?

I went to a night club one New Year's Eve with my dear friends, Burt and Nancy. It was a special night for us because the lead singer was a friend of Burt's and I had not seen Burt and Nancy in years. The band was well known, and the high-class, very expensive dinner club was packed with people eager to hear the band. To get more tables in the club, dinner tables were placed all the way up to the stage, eliminating the dance floor. As we sat and listened, I could feel the gyrations of the music running through me until it was impossible to keep my body from moving to the music while still in my seat. Looking around the club, I noticed some finger tapping and maybe a foot or two keeping step, but everyone was firmly planted in their seats. This made sense because if you wanted to dance you would literally have to weave between the tables. One false move would present the risk of falling and even the humiliation of enduring a face plant in my new red dress. I convinced myself the logical and prudent move was to stay put. As the next song began, even more spectacular than the first, I began to feel remorse, even sorry for myself. Here I am with my best girlfriend and her husband in a moment we may never share again, yet I'm choosing to sit it out because I'm afraid something will go wrong. Finally, operating from a belief that our embarrassment lasts but a moment in time, yet our regrets last a lifetime, I asked Nancy if she would dance with me. It was a challenge to dance between the tables until we carved out a quiet corner of the room as our own space. Soon we both were dancing up a storm, smiling and laughing the whole time. Before we knew it, Burt came to dance with us, too. At first, I thought we would not be able to find space for all of us to dance, but we somehow managed to fit in an open corner of the room. When the song ended, we sat at our table again laughing and breathless. This time I didn't even give a thought to who was watching. I allowed myself truly to enjoy every moment, laughing and smiling with Burt and Nancy. Our waitress, who was taking an order at a nearby table, joined us in the dance. It felt so good to draw her into our joy. In that moment I felt our joy was contagious, infecting the room until, one by one, the couples sitting around our table had joined us in the dance. No one was disgruntled that the waitress paused to enjoy the moment, and most took joy in our spectacle. At the end of the evening, two women who were seated near our table came over to thank us for inspiring them to get up and enjoy the music on this final day of the year. As we shuffled out of the club after midnight, we were all laughing and joking about the wonderful evening we had experienced dancing with our waitress and the strangers around us. That night I learned a lesson that seems so obvious now: If you want to inspire joy, passion, and a new experience, be the first example.

At this point in the book, you have read about the importance of leading the way, so I won't belabor the point again here. Simply put, remember the old adage: "Don't do what I do, do what I say," does not work in any corner of nature. People emulate their leader. Whether you are wearing the hat of mentor, coach, project manager, team member, sponsor, or CEO, you need to model the behavior you expect others to follow. There are no excuses and no exceptions to this one.

8.8 Understand What Makes Others Passionate

Passion is purpose. Understand this and you will comprehend why this step is the most important for promoting productivity and engagement on the team. Learn about what's important to the people that you spend more time with than you may with your own family. Whether we admit it or not, work relationships form our attitudes and impressions of the world around us, as much or perhaps more than the other relationships in our lives. Learn, grow, and expand your horizons through others around you. Build trust to go beyond the surface to a place of knowing and understanding the individuals on your team.

8.9 Turn Passion into Power

When it comes to passion, I have found that there are basically two distinct types of individual personalities: passionate and dispassionate. The labels of passionate or dispassionate are often applied to people based on their behavior and expressions. Determining whether people are passionate based on their expressions is where we need to apply caution. Things are not always as they appear when it comes to feelings of passion.

Dispassionate individuals are often labeled as cold and void of emotion. Being identified as dispassionate does not always carry a negative connotation, however. Appearing void of emotion can be a positive trait in high-stress, high-conflict situations and environments. For example, a police officer who came upon a murder scene and started bawling hysterically, or a social worker inspecting the home of an impoverished family while displaying a look of disgust, would not be viewed as professional or competent on the job. That is because, in stressful situations, we depend on police officers, social workers, women and men enlisted in military service, and others in public service positions to refrain from showing emotion. Controlling emotion in stressful situations often helps calm others and prevents escalating tensions by triggering an unwanted response. Does this mean people who are not displaying behavior typically associated with passionate individuals are not passionate?

Passionate individuals are typically identified by their behavior, which is often expressive and full of energy. These individuals are also characterized as people who are mission-driven and have a clear purpose which provides the compass and road map for their actions.

When it comes to determining whether someone is passionate about something, behaviors and expressions can be misleading. The trick to igniting each individual's passion on the project team is to read the signals appropriately and to dig beyond the surface.

Purpose is a powerful force that motivates individuals to accomplish remarkable things. Passion can ignite purpose, clarify direction, and make clear a course of action directed at fulfilling an individual's purpose. To tap into people's passion is to unleash a powerful force that will energize and motivate them to complete actions that align with their vision of personal and professional purpose and mission. Find it within each individual on your team, align it to the project mission and you have created a force that can make the impossible, possible.

8.10 Make it Safe to Fail, Safe to Learn, Safe to Share

Earlier in the book we talked about the need for each team member to feel safe within the team. This critical factor cannot be overstated as a key component of a productive, sustainable team environment. If you seek to develop the full potential of each individual on the team, a safe environment is a requirement, not a luxury. The leader is the individual largely responsible for creating a safe environment and works to create this through interactions with individuals on the team and within the group. As the leader, consider your actions to accomplish this, and choose them carefully so you validate the feelings and opinions of each individual on the team.

Remember, validating and acknowledging the contribution of ideas and opinions of others does not mean you are endorsing them. It simply means you are acknowledging the contributor of those ideas and opinions for their contribution to the group. The act of validation is especially important when someone on the team is openly sharing fears or concerns. Revealing these concerns and discussing them openly empowers you to work with the team to identify risks to success early enough to manage and perhaps eliminate the risk entirely.

Another, less widely recognized factor in a safe environment is the ability to assume positive intent. The ability to assume positive intent is created when you train yourself to take the position that those around you are motivated to produce positive results for and with others and driven to serve and support the well-being of the team, the project, and the organization. Assuming positive intent does not mean you are choosing to turn a blind eye to destructive behaviors and motives. It simply means you assume that the best intentions are driving the behavior of others, until you have facts that demonstrate beyond doubt that intentions are rooted in destructive or negative motives. People often assume negative intent to protect themselves from someone they believe may not have their best interest in mind. Too often, the practice of assuming negative intent becomes a self-fulfilling prophecy, quickly turning an environment into a toxic no-one-feels-safe zone.

If you want to create solid team performance, you will need to learn to control and eradicate the spoiler virus before it has a chance to spread. The way you lead a team will be a factor in creating an environment that discourages the spoiler virus, yet that alone will not be enough. You will need to learn to ignite the passion, that *raison d'étre,* within all members of the team so they not only understand and believe in the project mission, they aspire to realize the mission. Igniting this passion creates the thrill of the ride.

8.11 Finding the True Signals

In recent times our society has become more aware that people have trouble connecting, or even trusting, individuals who appear to be dispassionate.

At first, people might seem dispassionate until you strike the topic that inspires them. Although people are different in the degrees of passion they express, everyone has a fuse that is waiting for the right spark to ignite.

Some people are expressively passionate about an issue or topic. You can always tell, because whenever you bring up the topic, you will see a glimmer in their eye, then hear the pitch in their voice change, and maybe even see them break into an animated explanation of their view on the topic. We all know someone like this. In my family, it's my Auntie Bee.

Everyone who knows my Auntie Bee understands that bringing up the topic of politics is treading into potentially dangerous territory, because bringing up the topic will likely bring on a heated debate of some kind from which they will not be able to escape quietly. From Auntie Bee's behavior, which has become a predictable pattern over time, we believe Auntie Bee is very passionate about politics. We often joke about this. It's interesting that although Auntie Bee is very verbal and expressive about her views, she has never participated in an action to inspire political change, such as a demonstration or rally. For example, Auntie Bee has not volunteered on a political campaign committee or volunteered to serve as a pollster. Auntie Bee's passion might be easily sparked by a word or topic, but the spark is just not strong enough to motivate her to act to inspire changes that support her beliefs. It can be argued that because Auntie Bee does not become involved in actions to create change, politics is not really her passion, even though she sometimes becomes extremely animated when expressing her political views. Auntie Bee's passion is her purpose. Once you get to know Auntie Bee and establish an understanding of where she spends her time and what she does, you learn that her passion is the care and welfare of her husband, who is struggling with Alzheimer's disease.

Auntie Bee's days are consumed with the care of her husband and the hours of research she relentlessly pursues to identify new drug therapies, care remedies, and physical therapies that might improve her husband's quality of life. Auntie Bee's purpose is to lessen the progression of the disease to prolong their life together. Now, pause for a moment to imagine that Auntie Bee is a member of your project team. How would you identify her true passion and use it to serve the project and the team in a positive way?

People have many layers, and the reasons people do, or do not, act are complex and variable. To be an effective leader, you must refrain from judgment, look beyond the layers in each person, and work to understand the motivation for behaviors.

To find the passion in other people, you will need to start by showing genuine interest in getting to know them. You need to ask the right questions and demonstrate care and concern for their well-being. While you are not able to create people's passion, you can tap into their passion and help them use it to find purpose and meaning for the project and the team.

8.12 Is Seeing Believing?

Reflecting on the chameleon we discussed in Chapter 6, can you determine that people are dispassionate because you do not see a strong reaction from them in a situation where you would expect some type of emotional response? Perhaps these individuals appear to be dispassionate because they have mastered their chameleon, harnessing their behaviors to reveal only the emotion that will elicit the reaction they need from others, so they can navigate the demands of any situation effectively. The question prompts me to wonder whether a mastery of behaviors can cause us to assess someone as less trustworthy or less passionate or, on the flip side, can a mastery of behaviors make us more likely to trust someone we should not trust?

To find the answer, consider that we live in a world where seeing is believing. Because we assess any situation based on what we see first, dispassionate individuals who seem void of emotion in a situation are often viewed as less approachable or trustworthy than individuals who show emotion. This is because we interpret behavior as an indication of character. In fact, our interpretation of the behavior of another could be the exact opposite of a true indication of character. Flaws in our interpretations of what we see and hear may be one reason why fraud is so successful and prevalent in the United States and other countries. To examine this further, let's look at some statistics on fraud.

In 2015, True Link Financial reported that $36.48 billion in fraud occurs in the elderly community every year in the United States. A portion of this fraud, $9.85 billion, is attributed to successful con artist scams. This tells us that con artist scams contribute almost 30 percent of the total fraud committed against the elderly every year (2015).

Another, equally lucrative type of con artist scam is growing in the domain of online dating and romance sites. The Australian Competition & Consumer Commission (2018) reports that in 2017, over $81 million (AUD) in monetary losses were attributed to scams, with fraudulent investments being the highest contributor. Dating and romance scams led to the second highest losses in online fraud. The report further explained that the main tool con artists use to carry out scams in dating and romance, and scams targeting the elderly, are words and behaviors that trigger emotions and ignite passions in others that prompt them to act. Let's look at a typical example of a scam against the elderly to demonstrate this point.

8.13 The Power of Words

A con artist calls an elderly couple after learning on social media that their adult child is traveling in a foreign country. The con artist introduces himself as a friend of Billy, their son, who is currently traveling in South America. The con artist, let's call him Sam, goes on to explain to the couple that Billy has been arrested while they were hiking in the mountains. Sam has a photo of Billy and him hiking right before the arrest and offers to email it to the couple. When the couple receive the photo, they immediately believe Sam is doing everything he can to help Billy. Sam explains, to get Billy out of the local jail, he needs the couple to wire $10,000 immediately. The emotional tug is almost immediate for the couple, who would do anything to protect their child. When the emotional tug is strong enough, panic ensues, logic leaves the picture, and in our example, the couple wire the $10,000 to Sam, only to learn eventually that Billy is safe and sound and does not know anyone named Sam. Billy recognizes the photo as one he posted on Facebook, but the discovery comes too late. Billy's parents try to return Sam's call, only to find that the number is disconnected.

What do these examples tell us? Tap into someone else's passion, the thing that is most important to them, and you tap into the motive for a particular action or behavior. This is the key to triggering the emotion that will light the fuse for action.

8.14 The Making of a Thrill

I am one of those lucky people who has turned my passion for serving others and my passion for creating cool stuff with other people into a career that has lasted over 20 years. For me, there is no greater thrill than to inspire a group of people to make a difference by creating something that didn't exist before. I have never worked on a spaceship or designed a new kind of aircraft with a team. Yet when the team and I work together to meet a deadline and deliver a product that makes someone's life better, for me, there is no greater thrill. It's not about the type of product that we are developing, it's about the way we develop it because we develop it together. Throughout the journey we support each other, care about each other, help each other; we share, and we work together to overcome the obstacles and the emotions that we each face every day as individuals. The project is not a competitive environment where one individual rises above the others. It's an environment where each individual supports the others so we all feel important and cared for.

A few years ago, I was sitting one day on a bench next to the Danube River. It was the first time I had been to Hungary, so everything was new and magical. It was a sultry clear summer night with a million twinkling lights adorning the sky above me like diamonds on black velvet. The crescent moon was shining that night just enough to lay a cloak of light on the river in front of me. The scene was perfect in every way but one: I was alone. There was no one to share in the moment, and no one to witness the magnificence of the scenery or the adventure. The fact that I shared this moment with no one else but myself was bittersweet. In that moment, I learned the joy of an experience is diluted when it can't be shared.

Science tells us we are social creatures. We are designed to interact and share with others. Sharing both good and bad experiences intensifies the experience for the participants. This intensified experience is what happens on project teams. Emotions combined with the experience can amplify any situation in a group of people.

The experience on a project team works much the same way. The joy of the experience is greatest when it's experienced with others. The trick to making this happen is to identify what each person who is along for the ride views as a great experience, because we all have different ideas of what a great experience would be. As a leader, you need to find what each individual on your team views as a great experience. If you can identify and tap into something each member on the team is passionate about, you can inspire each member to create an amazing experience for the whole team.

8.15 Passion Is the Fuel for the Ride

Throughout the book, we've examined stories that cause us to reflect on typical situations that occur when people work together in teams or groups. While these stories may give us insight to understand some of the motivations for group and team behavior, the stories are not enough to create a roadmap that will tell us how to ignite the passion that lives within us all. Passion is a crucial factor if you want to unleash the true power of each individual within the group, inspiring innovation, and striving toward productive change. This is the type of passion that inspires action. It's a passion that inspires individuals to dedicate themselves to make a difference, inspires others to sit up and take notice, and motivates others to commit to a purpose beyond themselves to create meaningful, lasting change for us all. This type of passion fuels the ride that truly is the thrill of a lifetime.

Chapter 9

Teachers or Leaders—What's the Difference?

© Jim Kangas

Take a moment to think of what it was like on your first job. How did you learn what to do? Who taught you the ropes? Someone with the position of boss, trainer, or co-worker probably taught you a few things on that job. As you reflect on your experience, how would you classify the person who taught you the ropes? Did they fit the profile of a teacher, a mentor, a coach, or a leader? Regardless their job title, it's likely that your interactions with that individual fit one of those profiles. Whether you realized it at the time or not, the approach of the individual teaching you the ropes on your first job, their behavior and knowledge, determined whether your experience on the job was positive or negative.

The roles of leader and teacher are similar. Both roles influence us, lead us, and teach us. All leaders are teachers, even though all teachers may not be leaders. Leaders teach us by their actions and the positive or negative results they achieve.

As you have discovered throughout this book, when I refer to the role of a leader, I'm not referring to a position of power or a title; I'm referring to behavior. The role of teacher and leader is fulfilled through behavior that serves as an example. Knowledge is a factor in both leadership and teaching, of course, but if you want to positively influence outcomes, shape lives, and the world in which we live, you must live the example of what you hope to influence in others.

Maybe it was your boss or a co-worker who served as a mentor on your first job? Mentors are typically thought of as people who share their experience with another. A teacher is typically associated with someone who has expert knowledge of a specific subject area. Now think of your favorite teachers. What do you remember about your experience? When you think of your favorite teachers, past or present, what is the first thing that you remember about them? Do you first recall the classes taught by those teachers, or their character? Chances are you remember their characteristics as the factors that made your favorite teachers special to you.

The reality is that all of us are mentors, teachers, or students in some way. Sometimes we serve all three roles during an event or experience. It's simply a matter of acknowledging these roles and the fact that we serve them, consciously or not.

When we fulfill the role of mentor or teacher, we are stretching our skills and opening ourselves to learn along the way. To learn and to share knowledge is simple. It just requires a willingness to listen and help others.

Now reflect on your own behavior. Would you consider a mentor or teacher to be a leader? Before you answer, consider that it's often the great leaders throughout history who have taught us much about what we know of humanity. Martin Luther King, Jr., John F. Kennedy, and Nelson Mandela were teachers to us all through their leadership. The lessons they taught us still live on today.

We study great leaders of the past and present from around the globe to examine their wisdom, knowledge, and experience, in the hope of learning from their example. Whether we study the great teachers and leaders of history, or reflect on the example of our favorite teachers, we will find that great leaders and teachers all share the same traits of dedication, conviction, courage, and clarity in their mission to make a difference in the world. Each of us has the same opportunity to make a difference through our leadership. The opportunity to lead is not dependent on our socioeconomic position, race, creed, color, nationality, or even on our education level. Making a difference through leadership is a matter of mustering the sheer determination to create the legacy you envision through the life you chose to lead. It's a matter of choice.

To become a leader, the first step is to learn, and the journey does not stop there. In fact, the learning journey cannot stop if you intend to be an effective leader. To fulfill the role of leader, you must be committed to learning continuously from others and engage in the fluid, lively dance of learning, sharing, teaching, reflecting, and opening your mind to learning all over again. You must be willing to challenge your own beliefs and the beliefs of others. The ability to embrace humility as you let go of some of your own beliefs to accept new information is a building block that will empower you to sustain leadership that remains relevant through changing environments and cultures. The journey is one that requires an open mind, an open heart, and the courage to venture into the unknown. Are you ready?

9.1 Sharing Is Learning

A few years back, I received support from the Project Management Office (PMO) in my organization to start an internal Project Management Certification Program (PMCP) for individuals throughout the business who were not project managers or project coordinators. The goal was to help individuals throughout the organization gain a better understanding and appreciation for project management, the tools, the processes, and the value created as a result of adequate project leadership. As I began designing the program, I realized I would need to include a variety of viewpoints, styles, and perspectives on various project management experiences, tools, and techniques to develop a truly valuable program for participants. The best way to do this was to involve other project management professionals in the process of developing curriculum, hands-on exercises, and instruction.

One by one, the other project managers jumped onboard to sign up as instructors for the certification program. Finally, everything was in place one week before the program began, except for one lingering issue. Despite my best efforts, one project manager in the PMO was unwilling to support the PMCP. Although I tried to understand Tom's point of view, I just was not able to understand why Tom, as a Project Management Professional (PMP), was not willing to share his knowledge of project management. It was certainly true that we were all busy managing projects; even so, we managed to carve out time to participate as instructors and coaches. To ease the workload for the project managers participating in the certification program, we organized a small crew of support staff to help us design curriculum and tackle the administrative tasks that were needed to support the PMCP. As the program matured, we were able to train staff in the workforce to serve as project coordinators, who in turn supported project activities for the programs our PMPs managed. In this way, by working on an actual project, the learning investment paid off for the student and for the organization. To expose the program participants to the wide array of professional information, I shared resources from various project management sites, articles, blogs, webinars, and white papers.

Over time, I also came to understand that my role as instructor, coach, and mentor on the PMCP created a venue where I was required to reflect deeply and challenge my own decisions and their impact on a particular program or project. Reflecting continually on the methods I used, the impacts of my actions, and the results earned by my own behavior, positive or negative, created a deeper understanding of self, empowering me to find better ways of managing people, projects, and tools.

9.2 Teaching Tells Us Who We Are

In my quest to provide a wide selection of opinions, tools, techniques, and challenges to program participants, I soon found an unexpected challenge. As I reviewed new information, I discovered I needed to seek out and accept tools and approaches that were different than my own. This made me wonder: Had I become stuck in a project management rut? The fact that I found myself in this position was particularly odd because, up to this point in time, I had believed I was an agent of change. My personal mission was to stretch myself to embrace and adapt to change in any environment while remaining completely open to new ideas and different approaches. Now, I was beginning to think my own vision of self might be outdated, or worse, my belief that I was a champion of change might be wrong. Maybe I too was stuck in the quagmire of clinging to my own beliefs? This realization caused me to stop to reflect on who I really was and who I wanted to be. Once I committed to an honest, hard look at my own behavior and the results I was experiencing, it took me just a few minutes to conclude that I wanted to be the leader I believed I was—I wanted to be an agent of change. From that moment, my real journey began.

To help myself through the process of discovery, I created a method that would help me wade through the plethora of information on project management best practices. I identified the potential impacts of the practices, methodologies, and tools described in the information, matching them to the specific needs of my environment before I carried out an action or adopted a new approach. At first it was a nerve-racking process, because it was impossible to predict the outcome accurately when I tried something new. Through a series of trial and error, success and failure, I honed the process to a habitual practice. Over time, it became easier for me to critically analyze my actions and the tools I was applying, both before and after I tried something new. The process also made it easier for me to explain the value of my actions and help others take the same leap to consider adapting to a new idea, method, or behavior. In both the classroom and the project team room, I discovered that teaching and learning was paramount to building the skills in myself and in others to develop and expand new capabilities in project management, communication, and leadership.

9.3 Who Is the Real Mentor?

When I became confident as an instructor in project management, I thought I would branch out to expand my teaching and mentoring skills by signing up for a couple of mentoring programs through my local Project Management Institute (PMI) chapter. I was thrilled by the opportunity to share the journey of my growth and development with others. I believed that sharing my experience with mentees might help them reap the benefits of my experiences. It was simple, I thought, all I had to do was sign up to mentor someone and I would have the opportunity to help someone else through my experiences. I now understand how my belief at the time demonstrated how little I really understood about a mentor/mentee relationship. It would not take long until I discovered how wrong I was about who would be receiving the real benefits from this type of a relationship.

To begin my journey as a mentor, I signed up as a volunteer with an organization that connected military men and women who were transitioning out of the military to the world of project management in the civilian world. The organization matched each mentor and mentee based on the experience and goals of each participant. In this way, the process assured that the mentor and mentee were fairly aligned before their first introductory call.

The objective of the relationship was to assist the veteran in repackaging the valuable experience gained working on military projects for the civilian job market. The mentor process required an introductory call to learn about the mentee's goals and experience, while the mentor provided a short career and experience summary.

9.4 Bill, the Pilot

As I looked at my schedule for the day, I realized today is the day I spend my lunch hour on a call with Bill. All I knew about Bill was he was a military pilot, soon to retire to civilian life after spending 15 years in the U.S. Air Force. Bill was actively seeking employment in project management because he thought it would be the best fit for his skills. Bill was planning to move back to his hometown of La Crosse, Wisconsin, from his current post in Florida. So far, Bill was not having a lot of luck finding a position in an organization. From the information Bill provided me in preparation for the call, he had landed a couple of interviews but so far nothing had resulted in a follow-up interview or job offer.

I dialed into the conference line and began the call. "Hi, Bill, I'm Denise. I'm glad we were able to connect. How are you doing today?"

Bill sighed, "Ah, yes it's going OK, Denise. Thanks for agreeing to mentor me."

"Sure, Bill. Tell me, what I can do for you? What would you like out of this mentor/mentee relationship?" I asked.

"Well, Denise, I'm trying to figure out how to transfer my military experience organizing missions and teams on the flight line to civilian projects and organizations. Apparently, I'm not doing a very good job, because I'm not getting any follow-up interviews. I've applied for about 20 different project management positions, and each time I'm told I'm missing experience in the specific industry of the employer. It almost seems like no one wants a project manager to focus on project management. They want engineers, developers, or construction managers to do the job of engineering, developing, or construction while doing the job of project management too," said Bill.

I could hear the frustration in his voice, so I offered, "I understand how you feel, Bill. I've been there too. This town is a tough one to break into as well, especially when you're not already part of the community. I think I can offer some local connections that might be able to introduce you to some of the leaders in the community. Have you joined the local PMI chapter yet?"

As we continued the conversation, I realized I was learning from Bill every step of the way. Bill shared with me the tools and resources he was using to spruce up his résumé and refine his job search. As he did, I quickly understood that Bill was a wealth of information on the current tools and practices of job hunting. I, on the other hand, was out of touch on best practices for securing new employment because I had been out of the job search for quite some time. By the end of the conversation, we were both thanking each other for sharing information that broadened our horizons. As I disconnected the call, I thought to myself, "Wow, who knew I would learn so much from a mentor/mentee discussion? I almost feel like I was the mentee on that call. Bill has so much to offer."

For the next three months, Bill and I committed to a regular weekly check-in call. During these calls, I learned much more than I could have imagined from Bill's experiences, past and present.

Once Bill conquered the challenge of translating his experience in project leadership in the military to a story that could be easily understood by a local business, it did not take long for Bill to land his dream job.

Over time, as I continued to serve as a mentor, I realized I was the party receiving the greatest benefit from the mentor/mentee relationship. As a mentor, I was challenged to confirm my own knowledge and information, so I could be a credible resource for the mentee. At the same time, the mentee shared experiences and information with me, which delivered an unexpected treasure chest of new information, resources, and perspectives. The exchange broadened my perspectives, enriched my knowledge, and my experience. Every time I disconnected from a call with a mentee, I found myself pondering the conversation to recheck an opinion, position, or approach. The reward was a newfound ability to stretch my mind, my beliefs, and to remain open to unique ideas. This created a skill that paid dividends I could have never imagined before that first call with Bill.

To this day, I reserve time in my schedule to serve as a mentor, supporting project management mentor and coaching programs. Although I would still like to think I am sharing information that assists the mentee through the program, I know I am fooling myself if I think the mentee is the greatest benefactor in the relationship. Today, I now understand the time I invest to serve as a mentor for current or aspiring project management professionals is one of the most important steps I can take to continue to expand my own skills, broaden my horizons, and stay relevant. Now, whenever I sign up to engage in a mentor relationship, I wonder: Who will be the real mentor?

9.5 Project Managers Don't Teach, or Do They?

After the Project Management Certification Program had been underway for a year, I was asked to attend a PMO meeting to share my journey of instructing, mentoring, and learning with colleagues. During the meeting, I was more passionate than ever about the benefits of a PMCP. When I finished the presentation, I looked around the conference table to see nodding heads, confirming agreement with my message—everyone except for Tom. Tom sat still with a blank stare on his face. After a short pause, Tom spoke up saying, "We are project managers, not teachers."

Although I was a bit dumbfounded by this comment, I was able to gather my thoughts enough to say, "Tom, we are all teachers, regardless of our position or job title in the organization. We teach others and ourselves by what we say, by what we do, and by the choices we make. I believe it's simply a matter of opening our minds to the reality. It's our choice to apply our knowledge and experience in a way that creates a positive or negative influence over others. By giving yourself the opportunity to consciously make that choice, you pave the way to increase your own ability to lead others. What choice will you make, Tom?"

Tom shifted uncomfortably in his seat, and I realized I might have come on a little too strong. Tom looked down and the group looked at me in awkward silence. Then the meeting facilitator quickly thanked me for my presentation and shifted to the next agenda item. The PMO chapter meeting concluded just 15 minutes later.

As I went back to the table, I could not shake my feelings. I realized I was shocked and confused by Tom's comment. My perspective was so different, it was difficult to understand where Tom was coming from. During my career, I had found teaching was a key part of my work with every project team. I discovered during the initiation for every project that my ability to serve others adequately as a coach, mentor, and teacher was critical to helping project sponsors, product owners, team members, and other stakeholders understand the purpose and value of the tools, processes, expected outcomes, and required roles needed for the project. Then, once a project kicked off and the initiation phase was complete, it was often necessary to educate stakeholders on the process of identifying risks, risk triggers, owners, and their impacts on the project and the business. Methods that were needed to manage the project and keep it on track also needed to be explained to stakeholders and sponsors. If individuals on the project team had a different level of understanding of project methodologies, even more teaching and learning together on the project was required.

In addition to my own role as a teacher, it was critical for me to be a good student. In fact, everyone on the project team had roles as both teachers and students. These roles interchanged throughout the various phases of every project. When all members felt safe to share knowledge and expose their own knowledge gaps with the rest of the team, teaching and learning was a natural process because we recognized the project itself was a learning journey.

After the PMO meeting, as these thoughts ran through my mind, I realized the reason for my struggle. Projects in themselves are about learning and sharing knowledge. Each of us has something to share and so much to learn. I thought to myself, "How can we possibly embark on a journey to create a unique deliverable, a journey that has never been done exactly like this before, and not be open to learning, mentoring, and teaching each other?"

9.6 The Choice

The lessons we learn through the actions, failures, and successes of our leaders around the world are undeniable. If you serve your organization as a sponsor, product owner, portfolio, program, or project manager, you are already leading people. The questions remain how you will choose to lead, and what you will leave as your legacy of leadership.

Are you a teacher, a student, a mentor, or a mentee? If you have chosen to be a leader, a role synonymous with project manager, you have already accepted your responsibility to serve others as a teacher, a student, a coach, a mentor, and a mentee. There is just one question left: As a leader, how will you embrace your responsibility to serve others?

Chapter 10

What Will You Choose?

© Jim Kangas

Whether you are a project sponsor, just starting a career as a project manager, or you have worked in project management for years, you probably understand that the path to serving the project and the team to the best of your ability is full of twists and turns. The journey on a project is perilous because the world of projects and the profession of project management are more dynamic today than ever before.

Earlier in the book, we established your need to use a mix of tools, processes, and behaviors to navigate the journey. The key to success is to choose the right mix of tools, processes, and behaviors. As you put together the perfect mix and design your actions and behaviors, be mindful of the relevance of the information you use. I think we all would probably agree that basing choices on information that is 20 years old is a bad idea for any anyone. This is particularly true in project management, a profession designed to propel development into the future, create new products, and find new possibilities. So, if forward thinking is so important, why do we still base project management decisions on information that is 20 years old? I continue to ponder the possibilities. Maybe we stick to the old ways because change is just too hard, or maybe we lack the courage to challenge sponsors who resist change. Whatever the reason, it's important to remember that for every choice there is a benefit, an investment, and a cost. The benefit here is described as the value that can be realized as a result of the choice. The investment is the type of effort, knowledge, and behavior that is required to realize the benefit, and the cost is the potential effect, negative or positive, that results because of the choice.

Most of us are very familiar with the cost of change; I wonder, are we as familiar with the cost of stagnation?

10.1 The Cost of Change

If you've had an experience like Terry's in Chapter 2, you've probably been in a situation that clearly demonstrates the cost when a Project Management Office does not keep pace with the needs and perceptions of an organization.

Terry's story shows us that the cost of disbanding a PMO is wide reaching, impacting people, projects, the organization, its customers, and even the perception of the value of project management. With so much at stake, isn't the return worth your investment to become a change agent who is capable and willing to lead the kind of positive change that supports project success?

At this point, I'm guessing you understand that leading change in the project world requires first mastering and identifying the need for change, defining what the change should be, and finally applying the courage, discipline, and knowledge to implement change. Making this process even more of a challenge, the changes that are needed will be continuous, fluid, and dynamic. This same process of identifying change must also be applied to your personal attitude, behaviors, and professional abilities. If you enjoy a challenge, this is it.

10.2 Lessons from History

Through the example of Ferdinand de Lesseps in Chapter 1, we learned the importance of transparency and courage in the face of failure. We were able to see clearly the cost to a project, an organization, and a team when the project leader chooses not to meet the need to be transparent and courageous on a difficult project.

While on that fateful train trip in France, you had the opportunity to identify your own willingness to lead a team as an agent of change. On that trip, maybe you identified your resistance to change through the characterizations of a stick in the mud, or your unwillingness to go against the group as a cog in the wheel, or maybe you found your tendencies to behave like a steam roller, making impulsive decisions that drive change simply for the sake of change, as a skids on rails. What profile matches your reactions when faced with unpredictable events that cause changes to a plan and impacts the team?

In Chapter 3, we looked at the professional journey of Dr. Seuss and the indelible impact of President Lincoln's presidency in the United States, to examine the importance of courage and conviction when leading positive change and influencing others.

History, and a closer look in the mirror, should now give you insight into your own current capacity for change, whether you are called to adapt to changes or to lead them. Using lessons from history, whether the lesson is from a famous personality, or someone in our day-to-day lives, is an effective way to learn from someone else's mistakes to improve our own capacity for success. To succeed with this method, an investment is required. The process requires a solid knowledge of self, brutal honesty, an ability to keep an open mind and let go of old beliefs, and the willingness to change to go somewhere you have not yet ventured until now. If you can commit to the investment, the return will be greater than you can imagine.

10.3 A Matter of Evolution

If you still think change in project management can't possibly have an impact on the world like the kind made by Dr. Seuss, try to imagine whether Dr. Seuss felt he was changing the world of literature by writing a children's books.

Chances are, Dr. Seuss was also not able to assess the impact he would have on the world while he failed repeatedly to convince 27 publishers to take a chance on his unique approach, time and time again. Yet, through perseverance and conviction, Dr. Seuss did indeed change the way we all view the magic of literature. His example demonstrates that one individual choosing to commit to a different path can have a lasting impact throughout the world for generations to come.

We have already learned that failure is a part of the journey to success, so now you might be asking, "How many times can you expect to fail before you succeed?" To answer this question, let's examine the journey of the great inventor Thomas Edison. After inventing the lightbulb, Edison was asked how many times he had failed before he was finally able to get it right. To this question, the inventor responded brilliantly, "I have not failed, I just found 1,000 ways that won't work" (Edison, 2018).

Remember, conviction to lead change may not make you rich, or popular, and to be honest, it may even lead temporarily to the unemployment line, so why take the risk? The answer can be expressed in one word: evolution.

Think of an example of something that has gone extinct because it did not evolve quickly enough. There are plenty of examples in nature, the dinosaur and the wholly mammoth, to name only two.

Now, think of an example of a failure to evolve in business. What about the Polaroid camera, the Palm Pilot, the rotary dial phone, VCRs, and, of course, payphones? Every year, products and companies that do not keep up with the pace of change become obsolete and are forced to close their doors. As the pace of change becomes more rapid in our high-tech world, businesses and professionals who fail to create a culture designed to embrace change, inspire innovation, and challenge the status quo will experience the same fate as the dinosaur.

In nature and business, evolution is not a luxury; it's a requirement for survival. The best strategy is to accept that change is here. The question for you to answer is not whether you will deal with change, it's how you will deal with change. Do you choose to actively control your own destiny actively and lead change, will you choose to be passive and simply hang on for the ride, or will you choose to resist change? This choice may be one of the most important choices you make in your career, because it will determine whether you remain a viable survivor in your organization.

As you consider the choices in front of you, remember that resisting the wave of change is like resisting the power of a tsunami. Stand against a tsunami and you will be swallowed up in the wave. If you plan ahead for the tsunami, you will give yourself the opportunity to survive. Develop a method for rolling with the tsunami to stay above water and you give yourself the opportunity to thrive.

10.4 Develop a Plan

Just like my experience on the trip to Paris, we know that on any new journey, one person cannot make the trip a success for all. At the same time, we know that one person can ruin the experience for everyone. Your objective is to assure that you are not the person ruining the project experience.

Design your journey so you can demonstrate, through leadership, the qualities discussed in this book, to support projects and bring success to your team and your organization. Through this path, you will create the opportunity to build your own success, expand your future in project management, and add value to the profession.

To lead, you need first to know yourself. Break free of any notions you had regarding your effect on others and take a hard look in the mirror. Travel along the train trip in Chapter 1 and be brutally honest with how comfortable and how uncomfortable you feel in each scenario. Once you know who you are, you can define who you want to be. The gaps between who you are and who you choose to be in the future represent the goals for your own personal development plan. Design the plan to cover the action items that will be necessary to close the gaps in your capabilities. Understand who you are, make changes to become who you want to be, and work to define your legacy. This might sound easy, but believe me, it is not.

10.5 Model the Way

Although you can influence others and model the way, you can't control the actions and decisions of others. Keep your focus on understanding and not controlling the actions and perspective of those around you, so you can adjust your own behavior and actions to influence change in a direction that brings positive results for you, the team, and the project. This is a fluid continuous process that will require a substantial chunk of your time—that is, if you are truly devoted to modeling the way. Devotion and courage are required for this one.

10.6 The Right Message, the Right Time, the Right Method

In the workplace, people sometimes tend to hide emotional responses behind a myriad of facts, figures, and what can sound like erroneous rhetoric. The hard truth is that people make decisions on feelings and experience first and open their minds to consider new information last. Keep this hard truth in mind as you design project communication and choose the tools that will help you share the right message, at the right time, and using the right method.

Following this step does not mean you share only the good news. In fact, it cannot be overstated that sharing the good, the bad, and the ugly news on a project with expediency and accuracy is a crucial step in building your credibility and earning stakeholder confidence.

There is one more thing to include in this step: When you share communication with your stakeholders, speak in objective terms such as risk, reward, value, opportunity, and cost. Objective language will help you and your stakeholders tone down emotion, so you can focus on the issues and not the feelings. Stay mindful of the fact that feelings drive much of what goes on during a project. Why? Simply because humans are involved.

Remember those crucial components of cost, time, and scope that make up the components of the Iron Triangle? Develop communication to keep stakeholders informed on the current status of the components represented by the Iron Triangle.

Sharing the bad and ugly news on the project will require courage as well. Your ability to succeed in choosing the right message, at the right time, and using the right method for a particular audience will require a significant amount of emotional intelligence, EI, so brush up on those skills.

10.7 Know Where to Invest

When you or any participant on the project chooses to close the door to learning during the journey, the opportunity to succeed is over.

If you forget everything else from this book, remember this: Projects are not about the project plans they are written on. Projects are about the individuals and teams that lead them. Project plans are simply one of the many tools in your tool box. Factor in the importance of people and their behaviors in prioritizing the effort and time that you and the rest of the team invest in the project, every time. All the documents in the world will not build the deliverables for you, so you will need to create a positive experience, forge alliances, share knowledge, and create measurable value to deliver success. People do this. Teams do this. You do this.

Remember, when you run short on time and long on documentation, people create the successes or failures on projects, so invest your time in an approach that creates the value your team and your organization need. Invest your time where it matters, by developing a strong performing team.

10.8 Design How You Serve

Whether your role is project sponsor, project manager, or team member, approach your service on the project in a way that allows you to be the pedestal for the team, not the other way around. Don't forget that the power of your knowledge and experience originates from your ability to share it with others. Sharing your knowledge is how you will ultimately demonstrate your value.

Sharing and transparency are not processes that focus only on what was done right. Considering that we learn best by our failures, you will need to take a hard, honest look at what went wrong throughout the project journey. As you examine the facts, be genuine, honest, and transparent. Be the model to others for courage, trust, and transparency. Create a circle of safety for your team that makes it safe to fail, share lessons learned, and celebrate success. Inspire in others the commitment to try, the courage to fail, and the trust to share the journey along the way. This is your call as a leader, and it will require you to adopt and practice all the steps outlined in Chapter 4 to build and hone the necessary characteristics and behaviors of project leadership.

10.9 Look to the Future

Learn from the past, plan for the future, and keep your focus moving forward in a positive direction.

To a forward-thinking leader, the discussion is always less about what has been done and more about what will be done. While there is value in lessons learned, remember that success is earned by forward action.

Be the shield for your team, build trust, and earn respect. Protect your team from political quagmires, distractions, and barriers. Find out what each individual on your team needs to achieve optimum performance and work to provide it. If your team works hard, you will work harder. If you are thinking this is not the typical behavior of leaders in your organization, ask yourself: Do you want to be a typical leader, or do you aspire to be a great leader?

For every member of a team, success is not about focusing on the obstacles or opportunities as much as it is about who is standing beside you. Build your alliance with the team, and remember that project management is not a solo activity. You can't lead others who choose not to follow you. So, create the opportunities, inspire others to follow, and lead the way.

10.10 What Matters

Every day, every hour, we have a choice. Leadership is not a divine right for those only in the C-suite. Leadership is an opportunity for anyone who chooses to leave a legacy bigger than themselves.

Leaders in project management are essential, because these leaders pave the way for teams to create, innovate, and develop new possibilities, new opportunities, and even new worlds for all of humanity. As leaders, creating new worlds and opportunities is made possible by the choices each of us makes every day.

If you doubt you can make a difference that will leave an indelible mark on history, take a good look at the leaders we revere. These leaders were men and women who were not much different than you or me. The difference they made was possible because they made the choice to make a difference. It's not an easy choice, and it's not a choice you can make once and be done. This is a choice that you must continue to make every day, every hour. Along the way you will fail more than you will succeed, you will learn more than you can teach, and you will be challenged to the point of exhaustion. On this journey, perseverance and humility will lead to success, so be deliberate and continuous in your choice to lead. Do not succumb to weakness and self-doubt. Adopt and maintain a mission to serve, and cultivate your leadership path.

Looking back on my own journey through project leadership over the last two decades, I can clearly recall each time I was at a crossroads in my career. I could literally feel it in my gut. Each time, it felt like I was standing at the edge of a great divide, about to step onto a narrow wood-and-rope bridge. Once I crossed the bridge—if, that is, the bridge did not break and send me plummeting into the deep, dark, ominous valley below—I knew I would not be able to turn back and return to that familiar edge. Often, the choice to cross over the divide or stay on the edge was a choice that had to be made in an instant. If I didn't make this choice proactively, events would unfold that would determine my direction even if I didn't want to go anywhere! The journey was beyond my control, already in motion, and there was no way to stop it. Most of the time, when the moment occurred, I didn't want change in my life. I wanted to resist change and turn back time, so I could maintain my own status quo, stay comfortable, and veer away from the edge of that divide. It was a mind-blowing dilemma because I was terrified to step onto the bridge and terrified to stay on the edge. Yet all this was thrust upon me like a piano falling out of the sky in a single moment. In that moment, I knew that whether I chose to stay on the edge or walk over the divide on that shaky old bridge, my life would not be the same and the specific changes that would be unleashed would be impossible to predict. In the end, I almost always chose to take charge of the change and step off the edge. Sometimes, I only ventured a few steps out onto that bridge before experiencing my worst fears as the bridge collapsed underneath me. At other times, the bridge was incredibly strong, supporting each step forward. While the condition of the path forward was impossible to predict before I took that first step onto the bridge, I knew that if I remained committed to completing the journey, even if the bridge broke, I would figure it out and land on the other side. Each time I crossed the bridge, I learned a little bit more, increasing my ability to become more successful the next time I was standing on the edge of the divide, contemplating the decision to step onto that old bridge once again.

Your success in project leadership will probably not happen in a day, a week, or a month. This type of success may not even show measurable results in a year. Developing project management leadership skills and behaviors requires relentless effort every hour, day, week, month, and year. You may experience success in a single moment and minutes later experience failure. The key to success lies in your ability to persevere, to care for others, and your commitment to creating positive change. Continue the journey to become a leader and improve your capabilities daily. Make this choice and you will create your opportunity to succeed while having an unforgettable positive impact on everyone around you.

10.11 Putting It All Together

Projects and project management are about journeys into the unknown. To go on the adventure you will need tools, such as a map, and a vision of your destination which will serve as your compass.

You will not be on this journey alone. You will be the guide for a project team that will travel the road with you. Your mission is to build success for others.

To be the best guide possible, listen to advisors and the members of the team while you navigate the journey. Check your moral compass frequently, and be prepared for the challenge. Be compassionate and support those around you. Leave no one behind along the way, tuck everyone in at night, get them going with soul food in the morning, and keep them challenged while giving each of them permission to challenge you. Coach and cheer on each member in the group. Acknowledge the value and contribution of each individual and take time to enjoy the journey together. Remember to have fun frequently and to celebrate wins. Create the roadmap together. Track progress and failures so you can adjust the map throughout the trip. Be transparent, share, and communicate often to reduce fear and build confidence within the team and with others outside the team.

This journey requires you to lead, so listen to your gut, use, and build your knowledge, follow your heart, and look in the mirror often. When you look at yourself in the mirror, do you see someone who is a leader that others trust and would follow? If not, you might have some more work to do. The journey is not about the tools or the processes you use as much as it is about the people and their feelings, interpretations, cultural norms, expectations, and fears. The journey is about the human factor.

The human factor in project management is the key that brings success or failure to a project and a project team. The human factor in project management starts with you. Make no mistake: The choice to focus on the human factor is the most important choice you will make in your career. There is only one question remaining: What will you choose?

References

Chapter 1

CNN Library. (2013). "Los Angeles Riots Fast Facts." CNN. Retrieved May 12, 2017, from http://www.cnn.com/2013/09/18/us/los-angeles-riots-fast-facts/index.html

Federer, B. (2016). "High Price of Panama Canal, Then and Now." Retrieved February 19, 2018, from www.wnd.com/2016/02/high-price-of-panama-canal-then-and-now

Flow Psychology. (2014). "Mob Mentality Psychology." Retrieved May 12, 2017, from http://flowpsychology.com/mob-mentality-psychology

Global Security. (2011). "Panama Canal—Ferdinand de Lesseps 1880–1889." Retrieved May 12, 2017, from http://www.globalsecurity.org/military/facility/panama-canal-lesseps.htm

Goodreads. (2018). Retrieved May 21, 2018, from https://www.goodreads.com/quotes/search?utf8=%E2%9C%93&q=Progress+is+impossible+without+change&commit=Search

History.com Staff. (2009). "Rodney King Trial Verdict Announced." History. Retrieved November, 25, 2017, from http://www.history.com/this-day-in-history/rodney-king-trial-verdict-announced

Holt, J. (2018). Quotes. Retrieved May 12, 2017, from https://www.goodreads.com/quotes/49110-the-true-test-of-character-is-not-how-much-we

McLeod, S. (2017). "Maslow's Hierarchy of Needs." SimplyPsychology. Retrieved May 12, 2017, from https://www.simplypsychology.org/maslow.html

National Park Service. (2017). "Frequently Asked Questions (FAQs)—Flight 93 and September 11." NPS. Retrieved June, 10, 2017, from https://www.nps.gov/flni/learn/historyculture/sources-and-detailed-information.htm

Project Management Institute. (2017a). *A Guide to the Project Management Body of Knowledge (PMBOK® Guide)*, Sixth Edition. Newtown Square, PA, USA: Project Management Institute.

Project Management Institute (2017b). *PMI's Pulse of the Profession®, 9th Global Management Survey 2017*. PMI. Retrieved June 10, 2017, from https://www.pmi.org/-/media/pmi/documents/public/pdf/learning/thought-leadership/pulse/pulse-of-the-profession-2017.pdf

Chapter 2

Levin, G. (2016, September). "The Future of the PMO: Beyond Benefits and Value." PMI® Global Congress 2016—North America, San Diego, CA.
PM Solutions Research. (2014). "The State of the Project Management Office (PMO) 2014." PM Solutions. Retrieved May 12, 2017, from http://www.pmsolutions.com/reports/State_of_the_PMO_2014_Research_Report_FINAL.pdf

Chapter 3

Doyle, A. (2018, January 24). "How Often Do People Change Jobs?" The Balance. Retrieved May 12, 2017, from https://www.thebalance.com/how-often-do-people-change-jobs-2060467
Freeman, J. B., Stolier, R. M., Zachary, A. I., and Hehman, E. A. (2014, August 6). "Amygdala Responsivity to High-Level Social Information from Unseen Faces." *The Journal of Neuroscience, 34*(32):10573–10581. Retrieved November 25, 2017, from http://www.jneurosci.org/content/34/32/10573
History.com Staff. (2010). "President Lincoln Dies." History. Retrieved November, 25, 2017, from http://www.history.com/this-day-in-history/president-lincoln-dies
Kiisel, T. (2013, January 30). "82 Percent of People Don't Trust the Boss to Tell the Truth." Forbes. Retrieved October 11, 2017, from https://www.forbes.com/sites/tykiisel/2013/01/30/82-percent-of-people-dont-trust-the-boss-to-tell-the-truth/#287ef32c6025Be authentic
Kipman, S. "15 Highly Successful People Who Failed on Their Way to Success." Lifehack. Retrieved May 22, 2017, from http://www.lifehack.org/articles/productivity/15-highly-successful-people-who-failed-their-way-success.html
Markman, A. (2010, July 15). "Evaluating the Actions of Others." Psychology Today. Retrieved May 12, 2017, from https://www.psychologytoday.com/blog/ulterior-motives/201007/evaluating-the-actions-others
McKinley Irvin Family Law. (2012, October 30). "32 Shocking Divorce Statistics." McKinley Irvin Family Law. Retrieved May 22, 2017, from https://www.mckinleyirvin.com/Family-Law-Blog/2012/October/32-Shocking-Divorce-Statistics.aspx
Meek, W. (2013, July 13). "Basics of Communication." Retrieved February 24, 2018, from https://www.psychologytoday.com/blog/notes-self/201307/basics-communication
Merriam-Webster Dictionary. (2018, February 24). "Personality." Retrieved February 24, 2018, from https://www.merriam-webster.com/dictionary/personality
Mind Tools Content Team. (2018). "Plan-Do-Check-Act (PDCA)." Retrieved February 24, 2018, from https://www.mindtools.com/pages/article/newPPM_89.htm
PM Solutions Research. (2016). "The State of the Project Management Office (PMO) 2016." PM Solutions. Retrieved May 12, 2017, from http://www.pmsolutions.com/reports/State_of_the_PMO_2016_Research_Report.pdf
The Pulitzer Prizes. (2018). "The 1984 Pulitzer Prize Winner in Special Awards and Citations: Theodor Seuss Geisel (Dr. Seuss)." Retrieved May 12, 2017, from http://www.pulitzer.org/winners/theodor-seuss-geisel-dr-seuss
The Editors of *Encyclopedia Britannica*. (2018, February 8). "Richard Nixon. President of the United States." Britannica. Retrieved October, 14, 2017, from https://www.britannica.com/biography/Richard-Nixon

Tugend, A. (2012, March 23). "Praise Is Fleeting, but Brickbats We Recall." *The New York Times*. Retrieved November 21, 2017, from http://www.nytimes.com/2012/03/24/your-money/why-people-remember-negative-events-more-than-positive-ones.html

Chapter 4

Agile Manifesto. (2001). "Manifesto for Agile Software Development." Retrieved February 25, 2018, from http://agilemanifesto.org

Institute for Health and Human Potential. (2018). "Definition of Emotional Intelligence." IHHP. Retrieved November 28, 2017, from https://www.ihhp.com/meaning-of-emotional-intelligence

Project Management Institute (PMI). (2017a). *A Guide to the Project Management Body of Knowledge* (*PMBOK® Guide*), Sixth Edition. Newtown Square, PA, USA: Project Management Institute.

Project Management Institute. (2017b). *Agile Practice Guide*. Newtown Square, PA, USA: Project Management Institute.

Royce, W. (1970). "Managing the Development of Large Software Systems." Retrieved February 25, 2018, from http://leadinganswers.typepad.com/leading_answers/files/original_waterfall_paper_winston_royce.pdf

Tasler, N. (2016, June 22). "Why 1 in 3 People Adapt to Change More Successfully." Psychology Today. Retrieved October 22, 2017, from https://www.psychologytoday.com/blog/strategic-thinking/201606/why-1-in-3-people-adapt-change-more-successfully

U.S. Department of Defense. (1985). "Military Standard Defense System Software Development." Retrieved February 25, 2018, from http://www.product-lifecycle-management.com/download/DOD-STD-2167A.pdf

Chapter 5

Knapp, M. L., Hall, J. A., and Horgan, T. G. (2014). *Nonverbal Communication in Human Interaction*, 8th ed. Boston, MA, USA: Wadsworth Cengage Learning.

Project Management Institute (PMI). (2017). *Agile Practice Guide*. Newtown Square, PA, USA: Project Management Institute.

Weinschenk, S. (2012, September 18). "The True Cost of Multi-tasking." Psychology Today. Retrieved October, 12, 2017, from https://www.psychologytoday.com/blog/brain-wise/201209/the-true-cost-multi-tasking

Chapter 6

First Nations Pedagogy Online. (2009). "Talking Circles." First Nations Pedagogy. Retrieved October 2, 2017, from http://firstnationspedagogy.ca/circletalks.html

Project Management Institute (PMI). (2017). "Project Management Job Growth and Talent Gap." PMI. Retrieved October, 10, 2017, from https://www.pmi.org/-/media/pmi/documents/public/pdf/learning/job-growth-report.pdf

Smith, M. K. (2005). "Bruce W. Tuckman—Forming, Storming, Norming and Performing in Groups." *The Encyclopedia of Informal Education.* Retrieved November 7, 2017, from http://infed.org/mobi/bruce-w-tuckman-forming-storming-norming-and-performing-in-groups
Winch, G. (2015, February 3). "Can Leadership Be Learned or Are You Born with It?" Psychology Today. Retrieved October 2, 2017, from https://www.psychologytoday.com/blog/the-squeaky-wheel/201502/can-leadership-be-learned-or-are-you-born-it
World Economic Forum. (2016, January). "Executive Summary. The Future of Jobs. Employment, Skills and Workforce Strategy for the Fourth Industrial Revolution." World Economic Forum. Retrieved October 10, 2017, from http://www3.weforum.org/docs/WEF_FOJ_Executive_Summary_Jobs.pdf

Chapter 7

Blair, E. (2016, November 4). "Wisconsin Public Radio. The Real 'Hacksaw Ridge' Soldier Saved 75 Souls Without Ever Carrying a Gun." NPR. Retrieved October 17, 2017, from https://www.npr.org/2016/11/04/500548745/the-real-hacksaw-ridge-soldier-saved-75-souls-without-ever-carrying-a-gun
Indacochea, S. (2006, February 25). "Infanticide and Population Control." Swarthmore College Environmental Studies. Retrieved May 30, 2017, from http://fubini.swarthmore.edu/~ENVS2/S2006/sindaco 1/Infanticide.html
Project Management Institute (PMI). (2017). *A Guide to the Project Management Body of Knowledge (PMBOK® Guide)*, Sixth Edition. Newtown Square, PA, USA: Project Management Institute.
South Palomares, J. K., and Young, A. W. (2017, September 19). "Facial First Impressions of Partner Preference Traits." Sage Journals. Retrieved May 30, 2017, from http://journals.sagepub.com/doi/10.1177/1948550617732388
United States History. (2018). "Ohiyesa, or Charles A. Eastman." U-S-History. Retrieved February 24, 2018, from http://www.u-s-history.com/pages/h3898.html
Wargo, E. (2006, July). "How Many Seconds to a First Impression?" APS Observer. Retrieved November 11, 2017, from https://www.psychologicalscience.org/observer/how-many-seconds-to-a-first-impression

Chapter 8

Amara, P. (2010, May 29). "49 Minutes: The Time Each Day the Average Family Spends Together." Independent. Retrieved May 30, 2017, from http://www.independent.co.uk/news/uk/home-news/49-minutes-the-time-each-day-the-average-family-spends-together-1987035.html
Australian Competition & Consumer Commission (ACCC). (2018, January). "Dating & Romance." Scamwatch. Retrieved May 28, 2017, from https://www.scamwatch.gov.au/types-of-scams/dating-romance
Schulte, B. (2015, March 28). "Making Time for Kids? Study Says Quality Trumps Quantity." *The Washington Post.* Retrieved May 28, 2017, from https://www.washingtonpost.com/local/making-time-for-kids-study-says-quality-trumps-quantity/2015/03/28/10813192-d378-11e4-8fce-3941fc548f1c_story.html?utm_term=.03ad05f16362

True Link Financial. (2015). "Elder Financial Abuse Report." True Link Financial. Retrieved May 28, 2017, from https://www.truelinkfinancial.com/true-link-report-on-elder-financial-abuse-012815

Chapter 10

Edison, T. (n.d.). BrainyQuote. Retrieved March 3, 2018, from https://www.brainyquote.com/authors/thomas_a_edison

Index

C

D

H

I

J

K

L

R

S

T

U

V

W